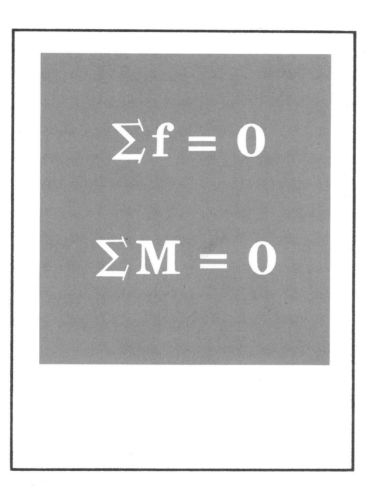

APPLIED MECHANICS:
STATICS

BOARD OF ADVISORS, ENGINEERING

APPLIED MECHANICS:
STATICS

CHARLES E. SMITH
OREGON STATE UNIVERSITY

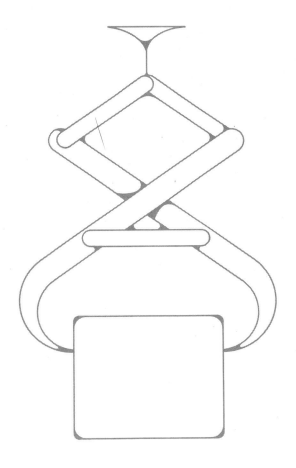

JOHN WILEY & SONS, INC.
New York London Sydney Toronto

This book was set in Baskerville by Progressive Typographers.
It was printed and bound by Halliday Lithograph. The drawings
were designed and executed by John Balbalis with the assistance
of the Wiley Illustration Department. The text and cover were
designed by Suzanne G. Bennett. The copyeditor was Deborah
Herbert. The production supervisor was Kenneth Ekkens.

Library of Congress Cataloging in Publication Data:

Smith, Charles Edward, 1932–
   Applied mechanics.

   Bibliography v. 1, p.   v. 2, p.
   Includes index.
   CONTENTS: v. 1. Statics   v. 2. Dynamics.
v. 3. More Dynamics.
   1. Mechanics, Applied.   I. Title.
TA350.S57   620.1   75-44021
ISBN 0-471-80460-6 (v. 1)

Printed in the United States of America

10 9 8 7 6 5 4 3 2 1

PREFACE

These three volumes on applied mechanics, *Statics, Dynamics,* and *More Dynamics* are designed to meet two needs. The first is for a textbook for the course in statics and dynamics, which is usually taught for students in several branches of engineering at the sophomore level. For students majoring in civil, chemical, electrical, and some other branches of engineering, instruction in these subjects usually ends after the equivalent of about six or eight quarter hours. The second need is for instruction in dynamics at the upper division level, for students who are specializing in mechanical, aerospace, and some types of "systems" engineering.

Although some present text books are satisfactory for the first course alone, transition into the second level, usually starting somewhere in the middle of another book, is awkward. Also, those books that are satisfactory for students who plan further study of dynamics are not written with the students of the first group in mind.

The first two volumes of this set, *Statics* and *Dynamics,* are written for a course that teaches students to apply the rudiments of mechanics and, simultaneously, provides the basis for study of the topics covered in the third volume. This volume, *More Dynamics,* covers subjects that are gen-

v

erally within the domain of mechanical, aerospace, and some "systems" engineers.

Concern for the needs of the student who will not study these latter topics dictates that the material offered in these volumes be organized somewhat differently than for a course that is designed only for mechanical and aerospace engineers. Specifically, treatments of kinematics and rigid body dynamics are each split, the more involved aspects, rotating reference frames and rigid body motions in which the direction of the angular velocity vector varies, are discussed in the third volume. Although this may not be the most efficient organization from a logical point of view, experienced teachers know that few students can handle this level of generality without some prior experience with the simpler special cases.

Trigonometry and first-year calculus are the required background for *Statics*. In Chapter 1, the rudiments of Newtonian mechanics are covered in sufficient detail that a course can be given without a physics prerequisite. If a physics prequisite is used, students can move more quickly through this material. Similarly, a class of students who have covered vector algebra can progress quickly through Chapter 3. As a minimum, a course in *Statics* should cover Chapters 1 through 4, Section 6-1, and enough of the remainder of the book to insure that students are adequately skilled in drawing free-body diagrams and in writing and solving equations of equilibrium. Other sections may be chosen nearly independently of one another, the choices depending on the objectives of the curriculum.

The heavy emphasis on the geometric interpretation of vectors reflects my belief that students cannot learn mechanics by replacing a visual process with formal manipulation of equations. Too often, attention is given to the formal algebra of vectors in terms of sets of components, at the expense of attention to the *meaning* of vector relationships basic to mechanics. Instruction should encourage students to seek a spatial understanding first and to refer to this constantly as analysis proceeds. For example, the symbol $\mathbf{A} \times \mathbf{B}$ should bring to mind immediately the sketch on page 38, but should suggest the formula (3-22) only when a computation is required.

Very few people can learn mechanics solely by observing the analysis carried out by someone else. At some point (the sooner the better) the student must discard the observer's role and, making and correcting the inevitable mistakes, attempt to work problems and carry out derivations with increasing independence. For this reason, these books will be relatively ineffective in the lap of someone sitting in an easy chair. They must be studied at a desk with an ample supply of scratch paper at hand. The paper will be needed for attempting solutions of example problems before given solutions are read, for carrying out steps of analysis that are omitted from the books, and for making supplemental sketches. A student who works in this way, always attempting to relate solution procedures to basic ideas, will achieve an understanding that is satisfying as well as professionally useful. In this endeavor I wish him well.

CHARLES E. SMITH

Corvallis, Oregon
July 1975

ACKNOWLEDGMENTS

There were many contributors to this project, and I thank all of them. Valuable suggestions came from many students, especially Brad Whiting and John Gale. Dr. Hans J. Dahlke pointed out a number of errors and shortcomings in *Statics*. Dr. William E. Holley, Dr. Robert W. Thresher, and Dr. Robert E. Wilson contributed significantly to *Dynamics*. I am grateful to the staff of Wiley for the editing and production. The illustrations were prepared by John Balbalis. The most eminent and a source of inspiration is Emeritus Professor Kenneth E. Bisshopp of Rensselaer Polytechnic Institute. Finally, for nearly all of the latter I owe thanks to Marian, Brian, and Susan.

C.E.S.

CONTENTS

1 Introductory Concepts   **1**

     1-1   Newton's Laws of Motion   2

     1-2   Newton's Law of Gravitation   5

     1-3   Dimensions and Units of Measurements   7

     1-4   Weight and Mass   11

2 Resolution of Force Systems   **13**

     2-1   Direction and Magnitude; Force Resultants   13

     2-2   Point of Application and Line of Action   20

     2-3   Moment of Force   21

     2-4   Couples   28

     2-5   Equivalent Force Systems   29

3 Vector Algebra   **33**

     3-1   Basic Concepts   33

     3-2   Addition and Subtraction   35

     3-3   Products   36

         Products with Scalars   36

Dot (or Scalar) Products   37
Cross (or Vector) Products   38

3-4 Rectangular Cartesian Components   41
Equations   41
Addition and Subtraction   42
Dot Product   42
Direction Cosines   44
Cross Product   45
Scalar Triple Product   45

3-5 Composition and Resolution of Force Systems   46
Moments of Forces   46
Varignon's Theorem   50
Equivalent Force Systems   51
Composition of Force Systems   52

4 Static Equilibrium   **59**
4-1 Free-Body Diagrams   60
What to Exclude and Include   61
Labeling   62

4-2 Equations of Equilibrium   68

4-3 Statically Indeterminate and Improper Support Systems   87
Redundant Supports   87
Deficient Supports   87
Criteria   87
Clamped Supports   88

5 Some Special Applications   **91**
5-1 Trusses   91
Notation   93
Equations from Joints   93
Equations from Sections   95

5-2 Couple-Supporting Members   103
Twisting and Bending Moments   104

5-3 Systems with Friction   108
Coulomb's Friction Law   108
Belt Friction   111

6 Distributed Forces   **119**
6-1 Single Force Equivalents   121
Center of Mass and Center of Gravity   121

Centroids

6-2 Submerged Bodies 129
Static Fluid Pressure 129
Force Resultants 132
Archimedes' Principle 134
Moment Resultants 136
Second Moments of Area 138
Parallel Axis Shift 139
Rotation of Axes 141
Mohr's Circle 142

6-3 Bending Members 148
Differential Equations of Equilibrium 148
Graphical Interpretations 151
Concentrated Forces and Couples 152

6-4 Flexible Lines 159
Horizontally Uniform Loading 161
Uniform Line Under Gravitational Forces 162
Useful Relationships of Catenaries 163
A Problem in Cable Laying 167

7 Virtual Work **173**
7-1 Work Done by a Force 173

7-2 Work Done by Forces on a Rigid Body 175

7-3 The Principle of Virtual Work 178
Equivalence Between the Principle of Virtual Work
and Vanishing Force Resultant 183

APPENDIX A Some Useful Numerical Values **189**
A-1 Physical Constants 189
A-2 Prefixes for SI Units 189
A-3 Units of Measurement 190

APPENDIX B Properties of Lines, Areas, and Volumes **191**
B-1 Lines 191
B-2 Plane Areas 192
B-3 Volumes 193

REFERENCES **195**

INDEX **197**

INTRODUCTORY
CONCEPTS

Exciting advancements with significant effects on methods of modern engineering are appearing at an ever-increasing rate. Improvements in electronic computers, new instruments for making measurements and analyzing the information that these computers produce, and many other innovations provide an effective set of tools with which engineers can carry out their tasks.

Yet, there is hardly a design that does not rely heavily on the relatively old science of *classical mechanics*. This view of the mechanical interactions among objects was first formalized by Sir Isaac Newton in his *Principia* in 1686. Important extensions of Newton's laws of motion were developed by D'Alembert (about 1743), Lagrange (about 1788), and Hamilton (about 1827). The subsequent mathematical developments of vector and tensor analysis, just prior to 1900, further facilitate our understanding and ability to use the principles of mechanics. The knowledge we convey here stems directly from these developments, so it is by no means new.

However, new and interesting *applications* for this science arise as rapidly as ideas for new devices, so that a working understanding of

1

Newton's laws and their extensions is as indispensable to the engineer as modern electronic devices and computer systems.

1-1

NEWTON'S LAWS OF MOTION. Newton's three laws were originally stated as follows.

1. Every body continues in its state of rest, or of uniform motion in a straight line, unless it is compelled to change that state by forces impressed on it.
2. The change of motion is proportional to the motive force impressed and is made in the direction in which that force is impressed.
3. To every action there is always opposed an equal reaction; or, the mutual actions of two bodies on each other are always equal and directed to contrary parts.

Before these notions can be used, the terms "uniform motion in a straight line," "change of motion," "motive force impressed," and "action and reaction" must be understood fairly precisely.

Toward such an understanding let us consider the rigid block depicted in Figure 1-1. The block—in this case the "body" referred to in Newton's laws—is free to slide on the perfectly smooth surface and is under the influence of the attached spring. If the spring is in its relaxed state, it will exert no influence on the block; then, if the block is at rest it will remain at rest, and if it is moving with some speed it will continue to move at this same speed. If the spring is held in an extended state, the rightward speed v of the block is observed to increase, the time rate of increase dv/dt depending explicitly and only on the extension in the spring. The influence that the spring exerts on the block is called the

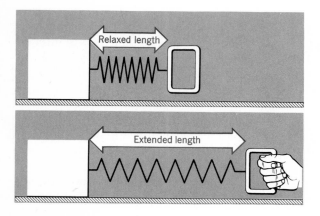

Figure 1-1

impressed *force,* and is directly associated with the observed time rate of change of speed, or *acceleration* of the block.

A quantitative definition of *force* is given operationally in terms of the observed acceleration of a standard object, a platinum cylinder located at the headquarters of the International Bureau of Weights and Measures in Paris. The force that will produce in this cylinder an acceleration dv/dt equal to A meters per second per second, is defined as having a magnitude of A *newtons.* The earth attracts a medium-small apple with a force of about one newton.

Now consider an extension of the above-described experiment. Suppose that for each of a number of different amounts of spring extension, we measure and record the acceleration of the standard object. We can now determine the force (originally defined in terms of the acceleration of the standard object) in terms of an observable quantity independent of the object, namely, the extension of the spring. Next, let us attach our newly calibrated spring to several different objects and measure the correspondence between force (now determined in terms of spring extension) and corresponding acceleration. The following fact emerges: for each object, the force f is directly proportional to the acceleration a; that is,

$$f = ma$$

The constant of proportionality m is a characteristic property of each object, called its *mass*, and is different for different objects. For example, the force required to impart a given acceleration to a hockey puck is considerably less than that required to impart the same acceleration to a goalie, which means that the mass of the puck is much smaller than that of the goalie. Or, a given force will impart a much greater acceleration to the puck than it will to the goalie.

Quantitative definition of mass is given in terms of that of the standard object discussed above. The *kilogram* is the unit defined as the mass of this standard object, which is called the international prototype of the kilogram. To determine the mass m of an object, we could compare its acceleration with that of the international prototype, both objects being subjected to the same force f. By using the subscript 0 to refer to the prototype, we can write

$$f = m_0 a_0 = ma$$

or

$$m = \frac{a_0}{a} m_0$$

In this way, we can determine the mass m of any object, in kilograms, as the ratio of the acceleration of the prototype to that of the other object.

The mass of a hockey puck is approximately 0.16 kg, and that of a suited-up goalie is approximately 110 kg.

In addition to magnitude, the concept of force must include consideration of direction. (Note that this word appears in Newton's statement of his second law.) We will tackle a precise meaning of direction as applied to velocity and acceleration in later chapters; for now, the direction of a constant force can be taken as the direction in which a rigid object, initially at rest and acted on by this and no other force, is observed to move.

As a means of depicting both magnitude and direction, we use an arrow, its length proportional to the *magnitude* of the force (i.e., the

number of newtons measuring it), and the direction indicated by its orientation with respect to the physical objects under analysis. Symbols used in equations and discussions to represent forces and other entities that have both magnitude and direction will appear throughout this book in boldface type, for example, **f**, **a**, and so forth. Thus Newton's second law, including its directional aspect, is written in the form:

$$\mathbf{f} = m\mathbf{a} \tag{1-1}$$

Newton's third law can be explained in terms of Figure 1-2. There, the block and spring have each been isolated from the objects with which they interact (i.e., the block isolated from the spring, and the spring isolated from the block and the hand), and the forces acting on each object are depicted by arrows. These diagrams are called *free-body diagrams*, and they are indispensable in the analysis of forces in mechanical systems. The force **f** from the spring acting on the block is directed to the right, causing the block to accelerate in that direction. Newton's third law states that the block must exert a force of equal magnitude and opposite direction on the spring. The spring is under the influence of this force −**f** from the block and the force exerted by the hand at the opposite end.

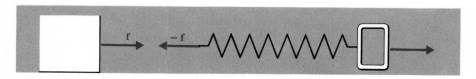

Figure 1-2

The force interaction in the example shown in Figures 1-1 and 1-2 occurs at the surface where the spring and block make contact. Force interactions can also occur between bodies that are not in contact but are merely in proximity with one another. Examples of this situation are electromagnetically induced forces and gravitational forces.

Problem

1-1   Explain how Newton's first *two* laws can be expressed by the single Equation (1-1).

1-2

NEWTON'S LAW OF GRAVITATION.   Another of Newton's great achievements was his deduction of the law that gives the magnitude of gravitational attraction between bodies. Together with his laws of motion, this law must be consistent with observed motions of the bodies in the solar system. It states that the magnitude of the gravitational force is given by

$$f_g = \frac{\gamma m_1 m_2}{r^2} \tag{1-2}$$

in which m_1 and m_2 are the masses of the two objects attracting one another, r is the distance between the centers* of the object, and γ is the *universal gravitational constant,* having the same value for all pairs of objects. When one of the two objects is the earth and the other object is near the surface of the earth (where r is about 6400 km), the group of factors $\gamma m_{earth}/r_{earth}^2$ is essentially constant, and we abbreviate it as g. Then the attractive law becomes

$$f_g = mg,$$

in which m is the mass of the object near the earth's surface. Precise measurements show a variation in g of about 0.5% among various locations on the earth's surface. For most estimates of gravitational forces near the surface of the earth, the value 9.81 N/kg is used. Thus a man having a mass of 82 kg will be attracted toward the earth by a force of magnitude

$$f_g = (82 \text{ kg})(9.81 \text{ N/kg}) = 804 \text{ N}$$

Consider now an object near the surface of the earth, that has no sig-

---

* The precise meaning of "center" in this case requires some fairly detailed analysis, which we postpone for now.

nificant forces acting on it other than the gravitational attraction exerted by the earth. Newton's second law of motion and law of gravitation give us the following value of the earthward acceleration of the body in this free-fall condition:

$$a_g = \frac{f_g}{m} = \frac{mg}{m} = g$$

This tells us that the acceleration is independent of the mass of the falling object. That is, every object will be accelerated at the same rate, regardless of its mass, if the earth's gravitational attraction is the only significant force. The fact that this agrees with observed behavior of free-falling objects provides one of the verifications of Newton's laws of motion and gravitation.

There are, of course, many other implications of Newton's laws of motion and gravitation. The exposition of some of them, and the development of skills that will enable the student to determine other implications that bear on the design of various mechanical devices, is the primary concern of this book.

Certain observed physical behavior cannot be satisfactorily predicted by Newtonian mechanics, most notably the behavior of particles of atomic size and objects traveling at speeds near that of light. Nevertheless, it is an amazing fact that, 300 years after their formulation, Newton's laws form the basis for successful prediction of the mechanical behavior of the vast majority of the systems that are employed by technology.

Problems

1-2   From the values of physical constants given in Appendix A-1, estimate the radius of the earth.

1-3   The radius of the moon is approximately 0.27 times that of the earth, and its mass is approximately 0.012 times that of the earth. At what rate will an object dropped near the surface of the moon be accelerated?

1-4   The mean distance between the earth and the moon is 384 400 km. Estimate the attractive force between them.

1-5   A space capsule is at a location on a direct line between the earth and the moon. The gravitational forces acting on the capsule exactly cancel each other. What is the altitude of the capsule?

1-6   A 45-Mg spaceship is at an altitude above the earth's surface of 200 km. Estimate the gravitational force. Estimate the acceleration of the *earth* caused by the force that the space ship exerts on the earth.

1–3

DIMENSIONS AND UNITS OF MEASUREMENTS.   Science and engineering are largely *quantitative* endeavors, so that they depend heavily on logical and consistent schemes for measuring physical entities and expressing interrelationships among measured values of them. We have just encountered the entities velocity, acceleration, force, and mass and have discussed ways of quantifying the last two. Any scheme of measurement is, ultimately, a set of comparisons with an agreed-on *standard,* such as the platinum cylinder discussed above. Standards for mass, length, and time, developed by the International Organization for Standardization (ISO) and accepted worldwide, are:

Mass. The *kilogram* is the unit of mass; it is equal to the mass of the international prototype of the kilogram. (An average man has a mass of about 75 kg.)

Length. The *meter* is the length equal to 1 650 763.73 wavelengths in vacuum of the radiation corresponding to the transition between levels $2p_{10}$ and $5d_5$ of the krypton-86 atom. (An average man is about 1.8 m tall.)

Time. The *second* is the duration of 9 192 631 770 periods of the radiation corresponding to the transition between the two hyperfine levels of the ground state of the cesium-133 atom. (This is the familiar one-sixtieth part of a minute.)

All other quantities in mechanics can be determined in terms of the quantities of mass, length, and time. For example, velocity is defined as an increment of distance traveled divided by the corresponding increment of time and, thus, can be measured in meters per second. The quantities mass, length, and time are said to be *fundamental,* and those of other mechanical quantities, such as velocity, are said to be derived.*

The *dimensions* of any physical quantity give the relationship that that entity bears with the quantities taken as fundamental. These dimensions are expressed in terms of M (indicating mass), L (indicating length), and T (indicating time). The dimensions of velocity, for example, are length divided by time, indicated as L/T. Acceleration is a change in velocity (L/T) divided by an increment of time (T), so that it has dimensions $(L/T)/T = L/T^2$ and can be measured in meters per second per second (m/s^2). Force is defined as the product of mass (M) and the acceleration it produces (L/T^2), so that it has dimensions ML/T^2 and can be measured in kilogram-meters per second per second (kg·m/s^2). This rather long unit

---

* In the U.S. engineering system of units, force (measured in pounds-force), length, and time are considered as fundamental quantities and the relationship $f = ma$ is used to derive mass. Thus the mass unit is the lbf·s²/ft, called the "slug."

of force has been given the shorter name "newton", and is abbreviated as N. That is, $1 \text{ N} = 1 \text{ kg·m/s}^2$. Dimensions and units of other quantities that are encountered in mechanics are given in the table appearing on the inside back cover and in Appendix A.

A bewildering number of systems of units are used in various disciplines and in different parts of the world. For many of them, the word system is overly flattering. In one office one might observe rates of energy transfer being calculated in watts by an electrical engineer, in horsepower by a mechanical designer, in British thermal units per hour by a heat transfer specialist, and in "tons of refrigeration" by a salesman of air conditioning units. The man-hours consumed by designers of refrigeration systems in making these units compatible would be appalling if known. Fortunately, a world-organized effort is underway to encourage those in all fields to adopt a single, coherent system of measurement units called "Le Systeme International d'Unites," or SI units. Conversion to the SI system is not likely to be completed for another decade or two, so that engineers will be expected to make correct conversions readily among various units for many years to come.

The best practice for handling this sometimes confusing task is to write the associated units (in abbreviated form) alongside each numerical value and then to cancel like unit symbols exactly as is done with algebraic symbols. To illustrate, let us suppose that the fuel economy of a new automobile is given as 21.3 miles per gallon, and that we need to know the equivalent in kilometers per liter. Reference to a table reveals that there are 5280 feet in one mile, 39.37 inches in one meter, 12 inches in one foot, and 3.7854 liters in one gallon. (If at this point you do not recognize the need for some systematic procedure for this type of calculation, it is suggested that you try to make the conversion before reading further.) A systematic method involves appropriately multiplying and dividing the given value by *unity*, unity taking forms like $1 = 5280 \text{ ft/1 mi}$, 12 in/1 ft, and so forth. Thus,

$$\text{fuel economy} = (21.3 \text{ mi/gal}) \frac{(5280 \text{ ft/mi})(12 \text{ in/ft})}{(39.37 \text{ in/m})(1000 \text{ m/km})(3.7854 \text{ liters/gal})}$$

Next, the like pairs of symbols representing various units are cancelled in the same manner as we cancel numerical quantities:

$$\text{fuel economy} = (21.3 \text{ m\!\!\!/i/ga\!\!\!/l}) \frac{(5280 \text{ f\!\!\!/t/m\!\!\!/i})(12 \text{ i\!\!\!/n/f\!\!\!/t})}{(39.37 \text{ i\!\!\!/n/m\!\!\!/})(1000 \text{ m\!\!\!/km})(3.7854 \text{ liters/ga\!\!\!/l})}$$

This done, we note that the remaining units are

$$\frac{1}{\text{liters/km}} = \text{km/liter}$$

as desired, indicating that we did not make a mistake such as multiplying by a conversion factor when division is correct. Finally, the arithmetic is done, yielding

$$\text{fuel economy} = 9.06 \text{ km/liter}$$

Of course, if one expects to do this conversion repeatedly, it would be wise to calculate and record the factor:

$$\frac{(5280 \text{ ft/mi})(12 \text{ in/ft})}{(39.37 \text{ in/m})(1000 \text{ m/km})(3.7854 \text{ liters/gal})} = 0.4251 \text{ km·gal/mi·liter}$$

Dimensional consistency must be satisfied by every equation used to describe physical phenomena. An equation is dimensionally consistent if every term has the same dimension. Use of this requirement is one of the most helpful ways of finding sources of errors in analysis of physical systems; therefore, the practice of making frequent dimensional checks is a sound habit. This checking can be accomplished by writing the dimensions of each quantity entering the equation, carrying out the multiplications and divisions indicated, and verifying consistency. For illustration, consider a formula that is to give the speed of fall of an object as it is pulled through a resistive medium by gravity. The analysis contains the assumption that the resistive force f_a is related to the speed v of the object as $f_a = cv$, where c is a constant depending on the viscosity of the medium and the size and shape of the object. The velocity of the object is supposed to be given as

$$v = \frac{mc}{g} \left[1 - e^{-(ct/m)} \right]$$

in which m is the mass of the object, g is the constant of gravity discussed previously, and t is time. To begin with, c is equal to f_a/v, so that it must have the dimension $(ML/T^2)/(L/T) = M/T$. Then, the argument ct/m of the exponential function is dimensionless $\{[(M/T)(T)]/(M) = 1\}$. This is as it must be, since the function $[1 - e^{-(ct/m)}]$ is expressible in terms of the series

$$(1 - e^{-\theta}) = \theta - \frac{\theta^2}{2!} + \frac{\theta^3}{3!} - \cdots$$

and the terms within this expression cannot be dimensionally consistent unless $\theta = ct/m$ is dimensionless. Next, the factor mc/g has dimensions $[(M)(M/T)]/(L/T^2) = M^2T/L$. But v has dimensions L/T, and so the equation cannot possibly be correct. The correct equation is

$$v = \frac{mg}{c} \left[1 - e^{-(ct/m)} \right]$$

which does pass dimensional inspection. Observe that satisfying dimensional consistency alone does not imply the validity of the relationship, but is necessary.

Problems

1-7   What is the fuel consumption, in liters per kilometer, of an automobile that gets 19 miles to the gallon?

1-8   Using values given in Appendix A, determine the following:
  The number of gallons in one ft^3.
  The density of water in lbm/ft^3.
  The speed of light in mi/hr.
  The number of feet in one nautical mile.
  The gravitational acceleration near the earth's surface, in ft/s^2.
  The density of water in g/cm^3.

1-9   Express the following in SI units.
  (a)   One light-year.
  (b)   A spring stiffness of 85 pounds-force per inch of displacement.
  (c)   The tension force in a cable from which a 10-ton block is suspended.
  (d)   The tension force in a cable from which a 2-Mg block is suspended.
  (e)   One kilowatt-hour.
  (f)   10 000 horsepower-hours.
  (g)   3000 feet per second.
  (h)   1700 rpm.
  (i)   4000 barrels per day.
  (j)   100 bushels per acre of wheat.
  (k)   1000 board-feet of lumber.
  (l)   48 000 acre-feet of water.
  (m)   Thermal conductivity of 127 Btu/hr·ft·°F.
  (n)   8729 furlongs per fortnight.
  (o)   "Speed 55 miles."
  (p)   "60 pounds of water pressure."
  (q)   "Six yards of gravel"

1-10   Check the following equations for dimensional consistency:

(a) $E = mc^2$ (Einstein's equation)

(b) $m \dfrac{d^2x}{dt^2} + kx = 0$

$\qquad x = x_0 \cos \sqrt{k/m}\ t$

m = mass
x = displacement
k = spring stiffness (force per unit displacement)

(c) $T = 2\pi \sqrt{a/g}$ T = period of oscillation of pendulum

a = pendulum length

(d) $f_c = mv^2/r$ f_c = centrifugal force

m = mass

v = speed

r = radius of curvature

p = fluid pressure

(e) $p = \rho g z$ ρ = density

z = depth

y = vertical distance to a point on a suspended cable

(f) $y = a \cosh x/a$

$a = S_0/\rho A g$ x = horizontal distance

S_0 = tension force at apex

ρ = density

A = cross-sectional area

(g) $A = \int_a^b y(x)dx$ A = area under curve

y = distance to curve

x = distance along x-axis

1–4

WEIGHT AND MASS. We now understand the mass of an object in terms of its resistance to acceleration, an "absolute" property, independent of the contact or proximity with other bodies. The technically correct meaning of the term *weight* is the force of gravity acting on an object, this force depending on the proximity of the object with a body such as the earth. Because it is possible to determine mass simply by measuring its weight (for instance, with a spring scale) the terms "weight" and "mass" are quite naturally confused. The confusion has been firmly entrenched by the unfortunate acceptance of the word "pound" for use as a unit of mass and as a unit of force in the U.S. customary system of units.

A 160-lb man is one who is attracted by the earth with a *force* of 160 lb when he is located near the surface of the earth. Unfortunately, it is accepted practice to also say that his mass is 160 lb. Measured with *consistent* units, his mass is expressed as

$$m = \frac{f_g}{g} = \frac{160 \text{ lb}}{32.2 \text{ ft/sec}^2} = 4.97 \text{ lb·sec}^2/\text{ft} = 4.97 \text{ sl}$$

Recall that with this system of units, force, length, and time are taken as fundamental quantities, and mass is derived through the relationship $f = ma$.

In the SI units of measurement, he would be referred to as a "73-kilogram man," meaning that his *mass* is equal to 73 kg. His *weight* (meaning, technically, the force of gravity) would be, near the surface of the earth,

$$f_g = mg = (73 \text{ kg})(9.8 \text{ m/s}^2) = 715 \text{ kg·m/s}^2 = 715 \text{ N}$$

Unfortunately again, his weight will generally be stated to be 73 kg, since it simply is not reasonable to expect most people to invest the necessary effort to understand the distinction between weight and mass. Therefore, the word "weight" will continue to mean both and is best avoided in technical work, where the distinction is important.

Problems

1-11   A spring scale and a balance scale, both of high precision, are calibrated at Washington, D.C. to measure mass. If they are taken to various parts of the world, what error can be expected from each?

1-12   The objects we see on television, floating about a space ship as it accelerates toward the earth, are commonly said to be weightless. Is this consistent with the technical definition of weight? State a definition of "weight" with which this usage would be consistent.

RESOLUTION OF FORCE SYSTEMS

2-1

DIRECTION AND MAGNITUDE; FORCE RESULTANTS. In analyzing forces it is necessary to account for both magnitude and direction. These aspects are depicted by arrows that have lengths proportional to the magnitudes represented. The magnitude and direction of an individual force is defined in terms of the acceleration that it will produce in a body in the absence of all other forces. When two forces act simultaneously, it is an observed fact that the acceleration produced, and hence by definition the resultant force, is determined by the *parallelogram law of addition* (Figure 2-1a, p. 14.) An equivalent procedure is the formation of triangles by the tail-to-head placements shown in Figure 2-1b and c. When any number of forces act, the resultant can be determined by repeated application of this law.

Example

What must be the tension in the line of tugboat *B* (Figure 2-2) so that the resultant towing force on the ship will be straight ahead? Also, what will be the magnitude of the resultant force?

13

Referring to the vector triangle we note that if the resultant is to be straight ahead,

$$T_B \sin 20° = (5 \text{ kN}) \sin 15°$$

Or,

$$T_B = \frac{(5 \text{ kN}) \sin 15°}{\sin 20°} = 3.78 \text{ kN}$$

Also from the vector triangle, we can determine the magnitude of the resultant as

$$f = 5 \text{ kN} \cos 15° + T_B \cos 20°$$
$$= 8.39 \text{ kN}$$

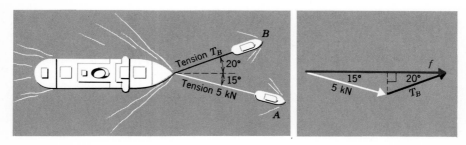

Figure 2-2

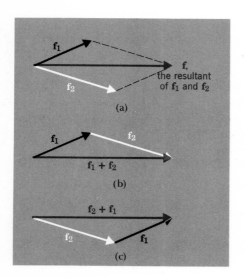

Figure 2-1

Problems

For Problems 2-1 through 2-6, determine the resultant force (magnitude and direction) of the given force systems.

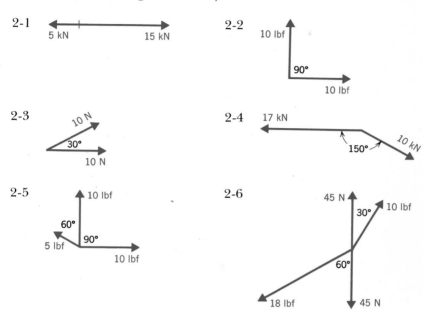

For Problems 2-7 through 2-12, determine the direction and magnitude of an unknown force **F** which, when added to the given force **G**, will yield the given resultant force **R**.

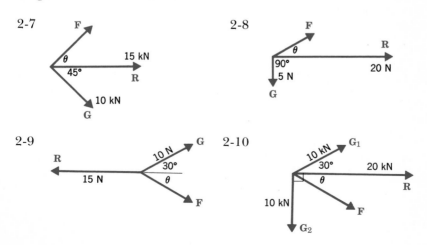

2-11

2-12

2-13   A ship is being towed by three tugboats as shown. Determine the resultant force exerted by the towlines on the ship. Each towline exerts a force of 5 kN.

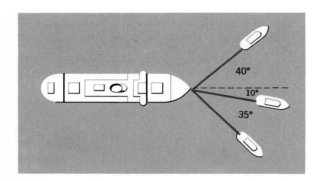

Resolution of force systems can be accomplished graphically by drawing a parallelogram as is shown in Figure 2-1 to scale. Resultant force magnitudes and directions can then be measured from the parallelogram. For Problems 2-14 through 2-17, determine the resultant force graphically.

2-14

2-15

2-16

2-17

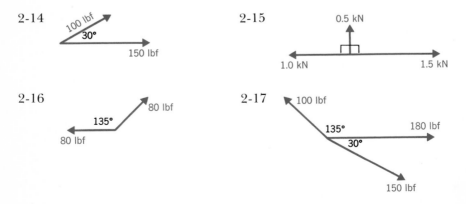

2-18   Determine the resultant force on the pillow bearing.

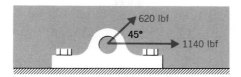

2-19   Determine the resultant force on the eyebolt.

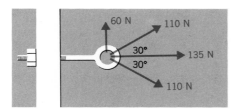

2-20   Determine the resultant force on the end of the diving board shown below.

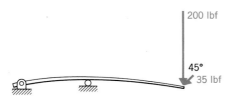

2-21   What is the resultant force on the skycycle?

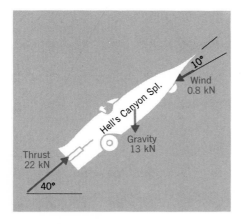

2-22   Determine the resultant force on the cover of your textbook as you open it.

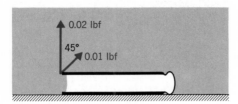

2-23   Determine the resultant force on the bracket.

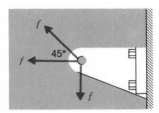

2-24   Spreader bars and stays are often used on vertical masts to increase the rigidity of the mast. If the tension in the upper portion of the cable is 600 N, what must be the tension in the lower portion so that the vertical components of the forces acting on the spreader bar at *A* cancel one another? What is the resultant force transmitted to the spreader bar by the cables?

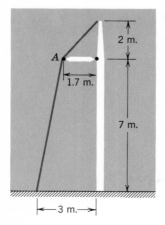

2-25 A rocket with 20 000-lb thrust is fired vertically. A horizontal wind blowing on the rocket tends to push the rocket sideways with a 2000-lb force. What is the resultant force on the rocket?

2-26 The contact point on a gear tooth has forces on it as shown below. Determine the resultant force on the gear tooth.

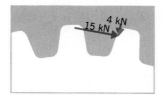

2-27 Determine the resultant of the three mutually orthogonal forces shown.

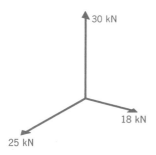

2-28 Determine the resultant force on the doorknob. The three components shown are mutually orthogonal.

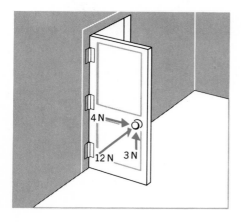

2-29 Determine the resultant force exerted on the tower by the two cables. The tension in each cable is 70 kN.

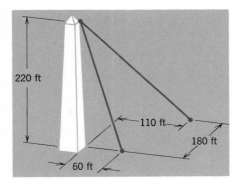

2-30 The tension in line *AC* is 450 lbf, and that in line *BC* is 1350 lbf. Determine the resultant force exerted by the two lines at *C*.

2–2

POINT OF APPLICATION AND LINE OF ACTION. The point of application of a force will in general have an effect on the way the body responds. Attaching a tow line to the front bumper of a stuck automobile might lead to removing the bumper from the automobile, whereas attaching it to a point on the frame might lead to the desired result. For many applications, however, the object may be idealized as *rigid;* that is, the deformation that the forces produce in the body can be assumed to be negligible. (The removal of a front bumper is not a negligible deformation.) Under this idealization, the observed effect of a force is the same for any point of application along a line parallel to the direction of the force. Such a line is called the *line of action* of the force. The fact that the effect of a force on a rigid body is the same for any point of application along the line of action of the force is called the *principle of transmissibility of forces*. A tugboat pushing on the stern of a steamship has the same effect on the steamship as one towing from the bow.

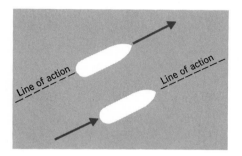

2–3

MOMENT OF FORCE. Anyone who has used a wrench is familiar with the "turning effect" that a force can have on an object. Furthermore, this "effect" is greater as the line of action is placed further from the point about which "turning" is considered. A more precise meaning of this intuitive notion is provided in the definition of *moment of force*.

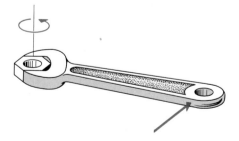

The moment about point O of the force $\mathbf{f}$ is defined in terms of the quantities shown in Figure 2-3. The magnitude of the moment is the product of the magnitude of the force and the perpendicular distance from O to the line of action of the force,

$$M_O = f \, \overline{OP}$$

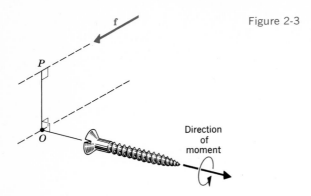

Figure 2-3

The *direction* of the moment is defined to coincide with the direction of advancement of a right-hand screw rotated by the force, the axis of the screw oriented perpendicular to the plane of the force and the point O.

Observe from the above relationship that the dimensions of moment of force are $(ML/T^2)(L) = ML^2/T^2$, and that it can be measured in newton-meters (N·m).

Keep in mind that the moment of a given force depends on the location of the reference point O. Thus it is ambiguous to refer to a moment unless the force, its line of action, *and* the reference about which the moment is reckoned are all specified.

Example

The friction in a pipe joint requires that a force of 220 N be applied at the end of a 0.2-m wrench handle in order to loosen it. What length handle is necessary so that the joint can be loosened by a 100-N force?

Assuming that the same moment about the joint will be required in both cases, we write this relationship in terms of the required handle length d:

$$M_O = (100 \text{ N})d = (220 \text{ N})(0.2 \text{ m})$$
$$(100 \text{ N})d = 44.0 \text{ N·m}$$

Then,

$$d = \frac{44.0 \text{ N·m}}{100 \text{ N}}$$
$$= 0.44 \text{ m}$$

This calculation was done with assumption that the wrench handle is perpendicular to the pipe axis and that the force is applied perpendicular to both the pipe axis and the wrench handle. How do you suppose the

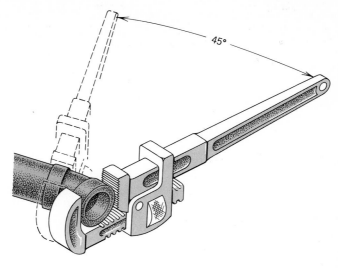

Figure 2-4

required force would change if the wrench were moved to the other end of a 45° elbow, as is indicated in Figure 2-4?

Consider next the moments about point O of the two mutually perpendicular forces $\mathbf{f}_1$ and $\mathbf{f}_2$ shown in Figure 2-5a, where the point O and the lines of action of the forces are all in the same plane. The directions of the moments are perpendicular to this plane, that of $\mathbf{f}_1$ directed into the paper (clockwise), and that of $\mathbf{f}_2$ directed out of the paper (coun-

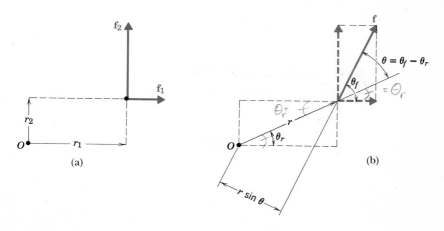

Figure 2-5

terclockwise). Let us define the sum of the moments with the convention that outward-directed (counterclockwise-acting) moments are positive and inward-acting (clockwise-acting) moments are negative. Then the sum of the moments has the value

$$M_{02} + M_{01} = r_1 f_2 - r_2 f_1 \tag{2-1}$$

Now consider the moment about O of the resultant force $\mathbf{f} = \mathbf{f}_1 + \mathbf{f}_2$ shown in Figure 2-5b.

$$
\begin{aligned}
M_O &= fr \sin \theta \\
&= fr \sin (\theta_f - \theta_r) \\
&= fr(\sin \theta_f \cos \theta_r - \cos \theta_f \sin \theta_r) \\
&= (r \cos \theta_r)(f \sin \theta_f) - (r \sin \theta_r)(f \cos \theta_f) \\
&= r_1 f_2 - r_2 f_1 \tag{2-2}
\end{aligned}
$$

Thus the sum of the moments of $\mathbf{f}_1$ and $\mathbf{f}_2$ is equal to the moment of the resultant $\mathbf{f}_1 + \mathbf{f}_2$! This is a special case of *Varignon's theorem*. Its generalization, to include any number of nonorthogonal forces in three dimensions, is easily established by using results from vector analysis, which is the subject of the next chapter.

Example

Evaluate the moment about O of the force acting on the end of the bracket (Figure 2-6).

The given force is equivalent to a downward force of 40 lbf, and a leftward force of 30 lbf. With the counterclockwise direction taken as positive, the moment is

$$
\begin{aligned}
M_O &= -(40 \text{ lbf})(9.8 \text{ in}) + (30 \text{ lbf})(7.1 \text{ in}) \\
&= -179 \text{ lbf·in}
\end{aligned}
$$

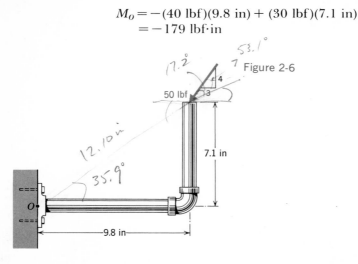

Figure 2-6

The minus sign indicates that the resultant moment about O is directed inward (clockwise).

Problems

2-31   The plumber exerts a force of 240 lb on the wrench handle. Determine the moment of this force (a) about point A, (b) about point B, and (c) about point C.

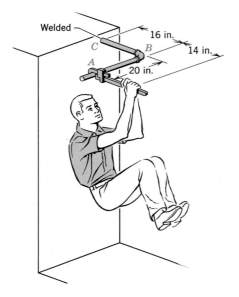

2-32   Determine the sum of the moments of the forces about the point C.

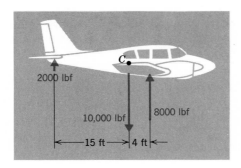

2-33   The moment about A of the force exerted on the springboard by the diver is 3.5 kN·m. What is the magnitude of the force?

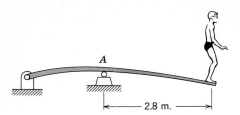

2-34   The connecting rod exerts a force of 1000 lbf on the crank as shown. Evaluate the moment about O of this force.

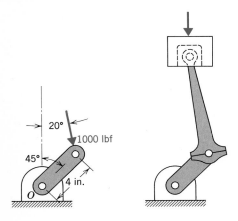

2-35   The signpost will break off at the ground if the sum of moments of forces reaches 1800 ft·lbf. Estimate the magnitude of wind force that the signpost can withstand.

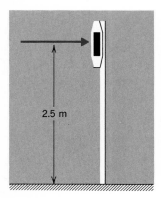

2-36   What will be the tension in the cable AB if the sum of moments about O, of the force this transmits to the boom and the 1000-lb force, is zero?

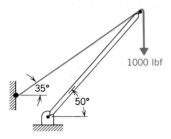

2-37   What will be the sum of the moments about O of the wind forces acting on the flag pole?

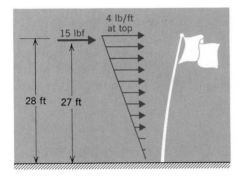

2-38   If the force from the nail is 1.5 kN, what must be the force P in order that the sum of moments about O of the forces acting on the crowbar is zero?

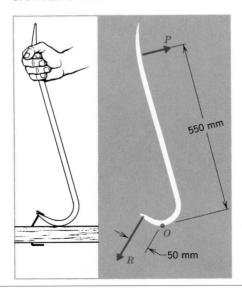

2-4

COUPLES. Two forces, of equal magnitude and opposite direction, and with different lines of action, constitute a force system called a *couple*. A couple has the properties that *its resultant force is zero, and its moment, about any point, has a magnitude equal to the product of the magnitude of either force and the perpendicular distance between the lines of action.* The first property is

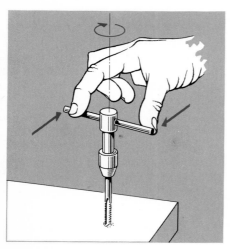

obvious; the second will be verified after we examine some helpful results from vector analysis. For now, let us examine the planar case where the point about which the resultant moment is evaluated lies in the plane of the lines of action (Figure 2-7). With the clockwise direction denoted by positive values, the sum of moments about O of the two forces has the magnitude

$$f(d + a) - fa = fd$$

which is independent of a.

Figure 2-7

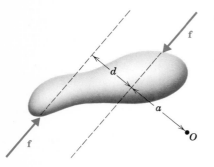

Problems

2-39   What is the direction of the moment, about the point where the tap wrench enters the block, of the two equal-magnitude forces that the operator is applying to the tap wrench depicted in the sketch on page 28?

2-40   If the operator of the tap wrench depicted in the sketch on page 28 applies a force from his thumb that is slightly greater than that applied by his finger, what will be the direction of the moment about the point where the tap enters the block?

2-41   Should the operator of the tap wrench depicted in the sketch on page 28 be sure that his thumb and forefinger are at equal distances from the vertical axis? Explain.

2-42   The brake is set on the wheel, and it will not slip until the moment about the center of the wheel of forces acting on the lug wrench reaches 150 N·m. Will the brake slip?

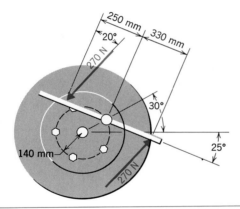

2-5

EQUIVALENT FORCE SYSTEMS. Two force systems are *equivalent* if they produce the same response in a rigid body. As we will learn later, this is the case provided that (1) the force resultants of the two systems are equal, and (2) the resultant moment about some point is the same for each system. For example, consider the systems of forces applied to the angle bracket in Figure 2-8. The resultant force in each case is 100 N downward, and the resultant moment about the corner of the bracket is 60 N·m directed outward from the wall, so that each system is equivalent to the others. What is the resultant moment of each system about the free end of the bracket?

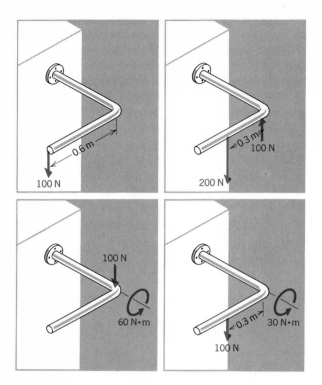

Figure 2-8

Straightforward methods of handling the details of various force systems can be developed by using the results of vector algebra, the subject of the next chapter.

Problems

Tugboats are busily maneuvering steamships about the harbor. For each steamship determine the single equivalent force and locate its line of action.

2-43

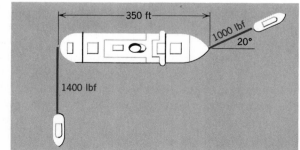

2-44

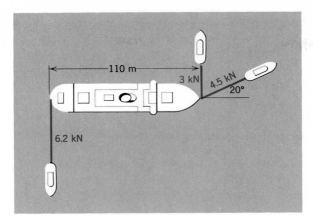

2-45

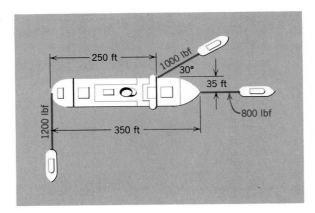

2-46 The bracket in Figure 2-8 has a force of 100 N downward, applied at a point 0.2 m toward the wall from the corner. Evaluate the additional couple that must be applied so that the force system is equivalent to the others shown.

2-47 The bracket in Figure 2-8 has a force of 100 N upward, applied at a point 0.2 m toward the wall from the corner. Determine a set of additional forces that, when added to this upward force, complete a force system equivalent to those shown.

VECTOR ALGEBRA

3–1

BASIC CONCEPTS. We call a quantity that has only magnitude (and, hence, can be represented completely by a single real number) a *scalar*. Examples are mass of a body and temperature at some location. A *vector* is a quantity that has magnitude and direction relative to some reference frame. Arrows are used to depict vectors. The length of an arrow is made proportional to the magnitude of the vector, and the direction is given by its orientation. Symbols used to represent vectors in this book will appear in boldface type, for instance, **A, B, a, b,** and so forth. The magnitude of a vector (a scalar) will be denoted either by vertical bars or by a corresponding letter without boldface, for example, $|\mathbf{A}|$ or A.

Nothing about location of a vector in the reference frame is implied in its specification. For example, the vector specifying the 42 km/hr southward velocity of a sportscar on East Elm Street is the vector equal of the 42 km/hr southward velocity of a limousine on West Walnut Street. Thus, if some physical quantity has, in addition to magnitude and direction, some "point of application" or "line of action," its analytical specification requires a *second* vector to locate this point or line.

The *projection* of a vector onto a line is the vector whose initial and terminal points are the projections of the initial and terminal points, respectively, of the given vector. We denote this with a subscript indicating the line onto which the projection is made, for example, the projection of $\mathbf{A}$ onto the line x is denoted by $\mathbf{A}_x$. The length of the projection is denoted as $|\mathbf{A}_x| = A_x$. It can be shown that the length of this projection is independent of the position, relative to the line, of the arrow representing $\mathbf{A}$.

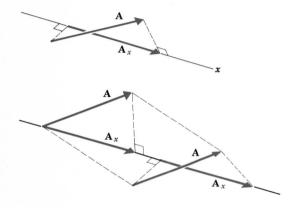

Problems

3-1

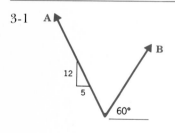

$$A = 65 \text{ ft/s}$$
$$B = 55 \text{ ft/s}$$
$$A_B =$$
$$B_A =$$

3-2 Show that the projection of a vector on a line is independent of the position relative to the line, of the arrow representing the vector,
    (a) For the case in which the line and the arrow are in the same plane, and
    (b) In general.

3-3 The value of a force may be completely defined by a vector, since this gives its magnitude and direction. Is the vector value of a force sufficient to determine the response it will effect on a body? Explain carefully.

3-2

ADDITION AND SUBTRACTION   The *sum* of two vectors is defined according to the *parallelogram law*. This is equivalent to the tail-to-head placement indicated in Figure 3-1. Successive application of the tail-to-

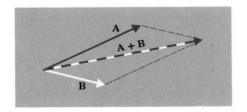

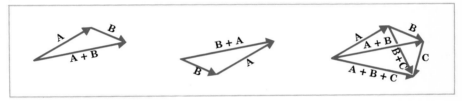

Figure 3-1

head rule gives the sum of several vectors. A brief consideration of the figures reveals that vector addition is

$$\text{commutative: } \mathbf{B} + \mathbf{A} = \mathbf{A} + \mathbf{B} \qquad (3\text{-}1)$$

and

$$\text{associative: } \mathbf{A} + (\mathbf{B} + \mathbf{C}) = (\mathbf{A} + \mathbf{B}) + \mathbf{C} \qquad (3\text{-}2)$$

From these laws, it follows that sums of any number of vectors can be formed, independent of the order of the tail-to-head placement.

The *negative* of a vector $\mathbf{A}$ is denoted by $-\mathbf{A}$ and is defined as the vector having a magnitude equal to that of $\mathbf{A}$ and direction opposite to that of $\mathbf{A}.$ Vector subtraction is then defined by

$$\mathbf{A} - \mathbf{B} = \mathbf{A} + (-\mathbf{B})$$

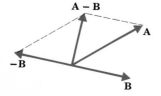

Problems

Determine the vector sum of forces in each set, and indicate whether the sets are equivalent with regard to response of a rigid body.

3-4

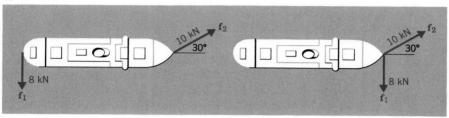

$$\mathbf{f_1} + \mathbf{f_2} = \begin{cases} \underline{\hspace{2cm}} \ \text{kN} \\ \underline{\hspace{2cm}}^\circ \ \text{from} \\ \text{ship axis} \end{cases}$$

$$\mathbf{f_1} + \mathbf{f_2} = \begin{cases} \underline{\hspace{2cm}} \ \text{kN} \\ \underline{\hspace{2cm}}^\circ \ \text{from} \\ \text{ship axis} \end{cases}$$

(a)   (b)

Equivalent? _____

3-5

$$\mathbf{A_1} + \mathbf{A_2} + \mathbf{f_g} \\ + \mathbf{T} + \mathbf{D} = \begin{cases} \underline{\hspace{2cm}} \ \text{lbf} \\ \underline{\hspace{2cm}}^\circ \ \text{from} \\ \text{horizontal} \end{cases}$$

(b) Same as (a), except the line of action of f_g is moved 2 ft to the rear.

$$\mathbf{A_1} + \mathbf{A_2} + \mathbf{f_g} + \mathbf{T} + \mathbf{D} = \begin{cases} \underline{\hspace{2cm}} \ \text{lbf} \\ \underline{\hspace{2cm}}^\circ \ \text{from} \\ \text{horizontal} \end{cases}$$

Equivalent? _____

3-3

PRODUCTS. We examine applications in mechanics for three kinds of multiplications with vectors, which are defined below.

Products with Scalars. Multiplication of a vector **A** by a scalar p results in a vector $p\mathbf{A}$ that has a magnitude pA and direction the same or opposite that of **A** according as p is positive or negative. The following rules result from this definition:

$$p(q\mathbf{A}) = (pq)\mathbf{A} \tag{3-3}$$

$$(p + q)\mathbf{A} = p\mathbf{A} + q\mathbf{A} \tag{3-4}$$

$$p(\mathbf{A} + \mathbf{B}) = p\mathbf{A} + p\mathbf{B} \tag{3-5}$$

Dot (or Scalar) Products. *Dot* (or *scalar*) multiplication of two vectors results in a scalar that is equal to the product of the magnitudes of the two vectors and the cosine of the angle between them. This is written as

$$\boxed{\mathbf{A} \cdot \mathbf{B} = AB \cos \sphericalangle {}^{\mathbf{B}}_{\mathbf{A}}} \tag{3-6}$$

Note that $B \cos \sphericalangle {}^{\mathbf{B}}_{\mathbf{A}}$ is the length of the projection of B onto a line in the direction of **A**. This is written as

$$B \cos \sphericalangle {}^{\mathbf{B}}_{\mathbf{A}} = B_A$$

Thus the dot product may be interpreted in terms of length of **A** and the length of the projection of **B** onto **A**:

$$\mathbf{A} \cdot \mathbf{B} = AB_A$$

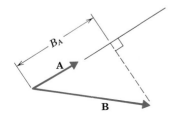

Furthermore, the important commutative property,

$$\mathbf{A} \cdot \mathbf{B} = \mathbf{B} \cdot \mathbf{A} \tag{3-7}$$

which follows from the definition (3-6), indicates that the projection interpretation may be made the other way around:

$$\mathbf{A} \cdot \mathbf{B} = A_B B$$

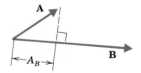

A consequence of the parallelogram law of addition is that the length of the projection of **A** + **B** on **C** is equal to the sum of the lengths of the projections of **A** on **C** and **B** on **C**:

$$|(\mathbf{A} + \mathbf{B})_C| = A_C + B_C$$

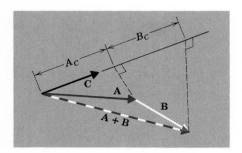

This in turn implies that the dot product is *distributive:*

$$(\mathbf{A} + \mathbf{B}) \cdot \mathbf{C} = \mathbf{A} \cdot \mathbf{C} + \mathbf{B} \cdot \mathbf{C} \tag{3-8}$$

Finally, it follows readily from the definitions that the dot product and multiplication by a scalar are *associative:*

$$(p\mathbf{A}) \cdot (q\mathbf{B}) = (pq)(\mathbf{A} \cdot \mathbf{B}) \tag{3-9}$$

Cross (or Vector) Products. Cross (or vector) multiplication of two vectors **A** and **B** results in a vector, written as **A** × **B**. This vector is defined to be perpendicular to **A** and **B**, to have magnitude equal to the

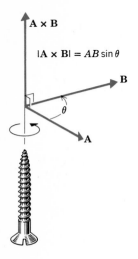

product of the magnitudes A and B and the sine of the angle between **A** and **B**, and to have direction determined by the right-hand rule; that is, its direction coincides with that of the advancement of a right-hand screw, the axis of the screw being oriented perpendicular to **A** and **B**, and turned in the direction **A** toward **B**.

Note that from this definition the cross product is *not* commutative; instead,

$$\mathbf{B} \times \mathbf{A} = -\mathbf{A} \times \mathbf{B} \tag{3-10}$$

However, as is indicated by the solution to Problem 3-19, it is *distributive:*

$$\mathbf{A} \times (\mathbf{B} + \mathbf{C}) = \mathbf{A} \times \mathbf{B} + \mathbf{A} \times \mathbf{C} \tag{3-11}$$

It can also be readily shown to be *associative* with scalar multiplication:

$$(p\mathbf{A}) \times (q\mathbf{B}) = (pq)\, \mathbf{A} \times \mathbf{B} \tag{3-12}$$

Other often useful relationships that are demonstrated in the exercise problems, are

$$\mathbf{A} \times (\mathbf{B} \times \mathbf{C}) = (\mathbf{C} \cdot \mathbf{A})\mathbf{B} - (\mathbf{A} \cdot \mathbf{B})\mathbf{C} \tag{3-13}$$

$$\mathbf{A} \cdot (\mathbf{B} \times \mathbf{C}) = \mathbf{B} \cdot (\mathbf{C} \times \mathbf{A}) = \mathbf{C} \cdot (\mathbf{A} \times \mathbf{B}) \tag{3-14}$$

$$\mathbf{A} = \frac{(\mathbf{B} \cdot \mathbf{A})\mathbf{B}}{\mathbf{B} \cdot \mathbf{B}} + \frac{(\mathbf{B} \times \mathbf{A}) \times \mathbf{B}}{\mathbf{B} \cdot \mathbf{B}} \tag{3-15}$$

Problems

3-6   Under what circumstances will **A** · **B** be negative? Under what circumstances will it be zero? Under what circumstances will **A** × **B** = **0**?

3-7   If **A** · **C** = **B** · **C**, does it follow that **A** = **B**? Explain.

3-8   If **A** × **C** = **B** × **C**, does it follow that **A** = **B**? If not, how are **A** and **B** related?

3-9   Given **B** and **C**, does the equation **C** = **A** × **B** uniquely define **A**? Explain.

3-10   Is **A** × (**B** × **C**) = (**A** × **B**) × **C**?

3-11   Show that **B** × (**A** × **B**) = (**B** × **A**) × **B**.

3-12   Show that the area A of the parallelogram with adjacent sides **a** and **b** is given by

$$A = |\mathbf{a} \times \mathbf{b}|$$

3-13   Show that $|\mathbf{A} \cdot (\mathbf{B} \times \mathbf{C})|$ is equal to the volume of the parallelepiped having **A**, **B**, and **C** as adjacent edges.

3-14 Show that the necessary and sufficient condition that **A**, **B**, and **C** all lie in parallel planes is that

$$\mathbf{A} \cdot (\mathbf{B} \times \mathbf{C}) = 0$$

3-15 The edges OP, OQ, and OR of the rectangular pyramid are mutually perpendicular. Show that the ratio of the area A_i normal to the i axis, to the area A of the triangle PQR, is equal to the direction cosine of the normal to PQR with the i axis, that is,

$$\frac{A_i}{A} = \cos \sphericalangle_i^{\text{normal}} \qquad (i = x, y, z)$$

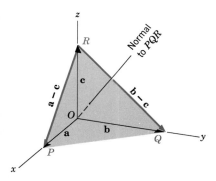

3-16 Derive Equation 3-13.

3-17 Derive Equation 3-14. *Suggestion:* refer to the result of Problem 3-13.

3-18 Explain in geometric terms the meaning of Equation 3-15. Verify its consistency with the identity (3-13).

3-19 (a) Show that if **A** is perpendicular to both **B** and **C**, then $\mathbf{A} \times (\mathbf{B} + \mathbf{C}) = \mathbf{A} \times \mathbf{B} + \mathbf{A} \times \mathbf{C}$.

(b) Show that if **B** is resolved into a component vector $\mathbf{B}_\parallel$, which is parallel to **A**, and a component vector $\mathbf{B}_\perp$, which is perpendicular to **A**, that $\mathbf{A} \times \mathbf{B} = \mathbf{A} \times \mathbf{B}_\perp$.

(c) Show that the distributive law (3-11) is valid in the general case, where there are no restrictions on **A**, **B**, and **C**. [Note that if **C** is resolved in the same way as **B** then, from the associative law for addition,

$$\mathbf{B} + \mathbf{C} = (\mathbf{B}_\parallel + \mathbf{B}_\perp) + (\mathbf{C}_\parallel + \mathbf{C}_\perp) = (\mathbf{B}_\parallel + \mathbf{C}_\parallel) + (\mathbf{B}_\perp + \mathbf{C}_\perp)$$
$$= (\mathbf{B} + \mathbf{C})_\parallel + (\mathbf{B} + \mathbf{C})_\perp]$$

3–4

RECTANGULAR CARTESIAN COMPONENTS. To *resolve* a vector is to replace it with two or more component vectors whose sum is the original vector. Of the infinitely many possible *resolutions* for a vector, perhaps the most useful is a set of three mutually perpendicular component vectors $\mathbf{A}_x$, $\mathbf{A}_y$, and $\mathbf{A}_z$. The equivalence is written as

$$\mathbf{A} = \mathbf{A}_x + \mathbf{A}_y + \mathbf{A}_z$$

in which the subscripts refer to a selected set of mutually perpendicular axes, as shown in Figure 3-2. It is also helpful to define a set of three

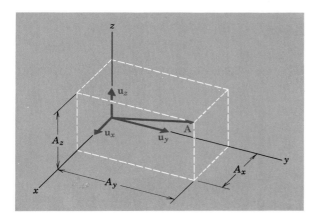

Figure 3-2

vectors, with unit magnitudes and in the directions of the component vectors, as

$$\mathbf{u}_x = \frac{\mathbf{A}_x}{A_x} \qquad \mathbf{u}_y = \frac{\mathbf{A}_y}{A_y} \qquad \mathbf{u}_z = \frac{\mathbf{A}_z}{A_z}$$

and to write the resolution of the vector as

$$\mathbf{A} = A_x \mathbf{u}_x + A_y \mathbf{u}_y + A_z \mathbf{u}_z \tag{3-16}$$

The $\mathbf{u}_i$ ($i = x, y, z$) are called the *unit base vectors* for this resolution, and the A_i are called the corresponding *components* of $\mathbf{A}$.

Equations. Analysis with vectors frequently leads to a result of the form

$$\mathbf{f} = \mathbf{g}$$

which may also be expressed as

$$f_x \mathbf{u}_x + f_y \mathbf{u}_y + f_z \mathbf{u}_z = g_x \mathbf{u}_x + g_y \mathbf{u}_y + g_z \mathbf{u}_z$$

Most applications require the use of the *component equivalent* of such a vector equation. This equivalent expresses the fact that the projections of two equal vectors onto the same axis are equal; that is, the above implies the three equations:

$$f_x = g_x$$
$$f_y = g_y$$
$$f_z = g_z$$

Addition and Subtraction.   Sums and differences of two or more vectors can be carried out in terms of the corresponding components in a set of Cartesian coordinate directions. From the laws (3-2) and (3-4),

$$\begin{aligned}
\mathbf{A} \pm \mathbf{B} \pm \mathbf{C} \cdots &= (A_x\mathbf{u}_x + A_y\mathbf{u}_y + A_z\mathbf{u}_z) \\
&\pm (B_x\mathbf{u}_x + B_y\mathbf{u}_y + B_z\mathbf{u}_z) \\
&\pm (C_x\mathbf{u}_x + c_y\mathbf{u}_y + C_z\mathbf{u}_z) \\
&\pm \cdots \\
&= (A_x \pm B_x \pm C_x \pm \cdots)\mathbf{u}_x \\
&+ (A_y \pm B_y \pm C_y \pm \cdots)\mathbf{u}_y \\
&+ (A_z \pm B_z \pm C_z \pm \cdots)\mathbf{u}_z
\end{aligned} \tag{3-17}$$

Thus the *i*th component of the sum or difference is equal to the sum or difference of the *i*th components.

Dot Product.   By using the expression (3-16) and algebraic properties established in the previous section, useful formulas for the dot and cross products can be derived. These formulas also depend on the following relationships, which are evident from Figure 3-2:

$$\mathbf{u}_x \cdot \mathbf{u}_x = \mathbf{u}_y \cdot \mathbf{u}_y = \mathbf{u}_z \cdot \mathbf{u}_z = 1 \tag{3-18a}$$

$$\mathbf{u}_x \cdot \mathbf{u}_y = \mathbf{u}_y \cdot \mathbf{u}_z = \mathbf{u}_z \cdot \mathbf{u}_x = 0 \tag{3-18b}$$

$$\mathbf{u}_x \times \mathbf{u}_x = \mathbf{u}_y \times \mathbf{u}_y = \mathbf{u}_z \times \mathbf{u}_z = \mathbf{0} \tag{3-19a}$$

$$\mathbf{u}_x \times \mathbf{u}_y = \mathbf{u}_z \qquad \mathbf{u}_y \times \mathbf{u}_z = \mathbf{u}_x \qquad \mathbf{u}_z \times \mathbf{u}_x = \mathbf{u}_y \tag{3-19b}$$

By using Equations 3-16, 3-8, 3-9, and 3-18, the dot product can be written as

$$\begin{aligned}
\mathbf{A} \cdot \mathbf{B} &= (A_x\mathbf{u}_x + A_y\mathbf{u}_y + A_z\mathbf{u}_z) \cdot (B_x\mathbf{u}_x + B_y\mathbf{u}_y + B_z\mathbf{u}_z) \\
&= A_xB_x\mathbf{u}_x \cdot \mathbf{u}_x + A_xB_y\mathbf{u}_x \cdot \mathbf{u}_y + A_xB_z\mathbf{u}_x \cdot \mathbf{u}_z \\
&+ A_yB_x\mathbf{u}_y \cdot \mathbf{u}_x + A_yB_y\mathbf{u}_y \cdot \mathbf{u}_y + A_yB_z\mathbf{u}_y \cdot \mathbf{u}_z \\
&+ A_zB_x\mathbf{u}_z \cdot \mathbf{u}_x + A_zB_y\mathbf{u}_z \cdot \mathbf{u}_y + A_zB_z\mathbf{u}_z \cdot \mathbf{u}_z
\end{aligned}$$

$$\boxed{\mathbf{A} \cdot \mathbf{B} = A_xB_x + A_yB_y + A_zB_z} \tag{3-20}$$

This formula has enough applications to warrant memorization. A special case of Equations 3-6 and 3-20 also has many applications. With $\mathbf{B} = \mathbf{A}$, they reduce to

$$\mathbf{A} \cdot \mathbf{A} = A^2 \cos 0$$
$$\mathbf{A} \cdot \mathbf{A} = A_x^2 + A_y^2 + A_z^2$$

Or,

$$\boxed{A = \sqrt{A_x^2 + A_y^2 + A_z^2}} \qquad (3\text{-}21)$$

This agrees with the result of applying the formula of Pythagoras to Figure 3-2.

Example

Determine the lengths of the guy lines OP and OQ, shown in Figure 3-3, and the angle between them.

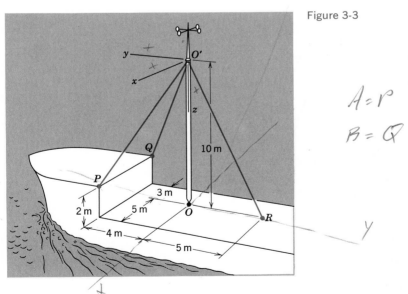

Figure 3-3

$A = P$

$B = Q$

Let position vectors $\mathbf{A}$ and $\mathbf{B}$ run from point O' to points P and Q, respectively, and resolve these into components along the x, y, z axes shown.

$$\mathbf{A} = (5 \text{ m})u_x + (4 \text{ m})\mathbf{u}_y + (8 \text{ m})\mathbf{u}_z$$
$$\mathbf{B} = (-3 \text{ m})\mathbf{u}_x + (4 \text{ m})\mathbf{u}_y + (8 \text{ m})\mathbf{u}_z$$

The required lengths can then be determined using Equation 3-21:

$$A = \sqrt{(5 \text{ m})^2 + (4 \text{ m})^2 + (8 \text{ m})^2} = 10.25 \text{ m}$$
$$B = \sqrt{(-3 \text{ m})^2 + (4 \text{ m})^2 + (8 \text{ m})^2} = 9.43 \text{ m}$$

Now, the required angle is a factor in the definition of $\mathbf{A} \cdot \mathbf{B}$, and in fact will be the only unknown in this equation after $\mathbf{A} \cdot \mathbf{B}$ is evaluated. From Equation 3-20 we have

$$\mathbf{A} \cdot \mathbf{B} = (5 \text{ m})(-3 \text{ m}) + (4 \text{ m})(4 \text{ m}) + (8 \text{ m})(8 \text{ m})$$
$$= 65.00 \text{ m}^2$$

Then the definition (3-6) gives

$$\cos \theta = \frac{\mathbf{A} \cdot \mathbf{B}}{AB}$$
$$= \frac{65.0 \text{ m}^2}{(10.25 \text{ m})(9.43 \text{ m})}$$
$$= 0.672$$

from which

$$\theta = 47.75°$$

Direction Cosines.   A useful way to specify *direction* relative to a set of rectangular Cartesian axes is to give the components of a unit vector in the desired direction. For example, we specify the direction of the vector $\mathbf{A}$ in Figure 3-2 by giving the components of the unit vector in the direction of $\mathbf{A}$:

$$\mathbf{u}_A = \frac{\mathbf{A}}{A}$$
$$= \frac{1}{A}(A_x\mathbf{u}_x + A_y\mathbf{u}_y + A_z\mathbf{u}_z)$$
$$= \frac{A_x}{A}\mathbf{u}_x + \frac{A_y}{A}\mathbf{u}_y + \frac{A_z}{A}\mathbf{u}_z$$

Observe that the magnitude of the ith component A_i/A of this unit vector is the length of the projection of the unit vector onto the i axis; therefore,

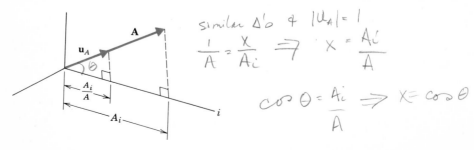

it is equal to the cosine of the angle between **A** and the i axis. The quantities A_x/A, A_y/A, and A_z/A are called the *direction cosines*, which specify the direction of **A**.

Example

Determine the direction cosines of the guy lines $O'P$ and $O'Q$ relative to the rectangular Cartesian axes shown in Figure 3-3.

   The length of line $O'P$ is $A = 10.247$ m. Then

$$\mathbf{u}_A = \frac{1}{10.247 \text{ m}} \left[(5 \text{ m})\mathbf{u}_x + (4 \text{ m})\mathbf{u}_y + (8 \text{ m})\mathbf{u}_z \right]$$
$$= 0.4880 \; \mathbf{u}_x + 0.3904 \; \mathbf{u}_y + 0.7807 \; \mathbf{u}_z$$

The length of line $O'Q$ is $B = 9.434$ m. Then,

$$\mathbf{u}_B = \frac{1}{9.343 \text{ m}} \left[(-3 \text{ m})\mathbf{u}_x + (4 \text{ m})\mathbf{u}_y + (8 \text{ m})\mathbf{u}_z \right]$$
$$= -0.3180 \; \mathbf{u}_x + 0.4240 \; \mathbf{u}_y + 0.8480 \; \mathbf{u}_z$$

Thus the direction cosines are

$$O'P: \quad (0.4880, \quad 0.3904, \quad 0.7807)$$
$$O'Q: \quad (-0.3180, \quad 0.4240, \quad 0.8480)$$

Cross Product.   Steps similar to those that led to Equation (3-20) lead to the following formula for the cross product.

$$\boxed{\begin{aligned} \mathbf{A} \times \mathbf{B} = (A_y B_z - A_z B_y)\mathbf{u}_x \\ + (A_z B_x - A_x B_z)\mathbf{u}_y \\ + (A_x B_y - A_y B_x)\mathbf{u}_z \end{aligned}} \tag{3-22}$$

It is important to realize that this result depends on the coordinate directions being *right-handed*, that is, on Equations 3-19 being satisfied. As an aid to memorizing Equation 3-22, observe that groups of subscripts in the order ⟳ are associated with positive terms and that those with subscripts in the order ⟳ are associated with negative terms.

Scalar Triple Product.   The product indicated in Equation 3-14 may be computed in terms of rectangular Cartesian components by combining Equations 3-20 and 3-22. A convenient form of the result is the determinant

$$\mathbf{A} \cdot (\mathbf{B} \times \mathbf{C}) = \begin{vmatrix} A_x & A_y & A_z \\ B_x & B_y & B_z \\ C_x & C_y & C_z \end{vmatrix} \tag{3-23}$$

Problems

For Problems 3-20 through 3-25, evaluate the unit vector in the direction of $\mathbf{A}$, $\mathbf{A} \cdot \mathbf{B}$, the angle between $\mathbf{A}$ and $\mathbf{B}$, $\mathbf{A} \times \mathbf{B}$, $\mathbf{A} \cdot (\mathbf{B} \times \mathbf{C})$, $\mathbf{A} \times (\mathbf{B} \times \mathbf{C})$, the components of $\mathbf{A}$ parallel to $\mathbf{B}$ and perpendicular to $\mathbf{B}$, $\mathbf{A} \cdot (\mathbf{B} + \mathbf{C})$, and $\mathbf{A} \times (\mathbf{C} - \mathbf{B})$.

3-20 $\mathbf{A} = 3\mathbf{u}_x + \mathbf{u}_y - 2\mathbf{u}_z$
$\mathbf{B} = 2\mathbf{u}_x + 2\mathbf{u}_y + \mathbf{u}_z$
$\mathbf{C} = \mathbf{u}_x + \mathbf{u}_y - \mathbf{u}_z$

3-21 $\mathbf{A} = \frac{3}{13}\mathbf{u}_x + \frac{4}{13}\mathbf{u}_y - \frac{12}{13}\mathbf{u}_z$
$\mathbf{B} = \mathbf{u}_x - 2\mathbf{u}_y - \mathbf{u}_z$
$\mathbf{C} = -2\mathbf{u}_x + 4\mathbf{u}_y + 2\mathbf{u}_z$

3-22 $\mathbf{A} = -\mathbf{u}_x + \mathbf{u}_y + 2\mathbf{u}_z$
$\mathbf{B} = \mathbf{u}_x + 3\mathbf{u}_y - \mathbf{u}_z$
$\mathbf{C} = -\mathbf{u}_x - 3\mathbf{u}_y + \mathbf{u}_z$

3-23 $\mathbf{A} = \frac{7}{25}\mathbf{u}_x - \frac{24}{25}\mathbf{u}_y$
$\mathbf{B} = \mathbf{u}_x - \mathbf{u}_y + \mathbf{u}_z$
$\mathbf{C} = \mathbf{u}_x + \mathbf{u}_y + \mathbf{u}_z$

3-24 $\mathbf{A} = \mathbf{u}_x + \mathbf{u}_y + \mathbf{u}_z$
$\mathbf{B} = \mathbf{u}_x + \mathbf{u}_y - \mathbf{u}_z$
$\mathbf{C} = \mathbf{u}_x - \mathbf{u}_y - \mathbf{u}_z$

3-25 $\mathbf{A} = \frac{1}{2}\mathbf{u}_x - \frac{1}{2}\mathbf{u}_y + \frac{\sqrt{2}}{2}\mathbf{u}_z$

$\mathbf{B} = \frac{\sqrt{2}}{2}\mathbf{u}_x + \frac{\sqrt{2}}{2}\mathbf{u}_y$

$\mathbf{C} = -\frac{1}{2}\mathbf{u}_x + \frac{1}{2}\mathbf{u}_y + \frac{\sqrt{2}}{2}\mathbf{u}_z$

3-26 Verify that Equation 3-20 is consistent with Equation 3-7.

3-27 Derive Equation 3-22.

3-28 Verify that Equation 3-22 is consistent with Equation 3-10.

3-29 Verify the consistency of Equations 3-20 and 3-22 with Equations
(a) (3-13),
(b) (3-14),
(c) (3-15).

3-30 Sketch a left-handed triad of unit base vectors and derive the corresponding formula for the cross product in terms of the rectangular Cartesian components. Compare the result with Equation 3-22.

3–5

COMPOSITION AND RESOLUTION OF FORCE SYSTEMS.

Moments of Forces. The most important application we have for the cross product is the evaluation of *moments*. A moment of a force is a vector with magnitude and direction as defined in Section 2-3. Examination of the definitions of moment of force and the cross product reveals that *the moment about a point O of the force* $\mathbf{f}$ *is given by*

$$\mathbf{M}_O = \mathbf{r} \times \mathbf{f} \tag{3-24}$$

where $\mathbf{r}$ *is a position vector from O to any point on the line of action of* $\mathbf{f}$.

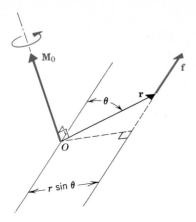

Example

Suppose the guy line $O'P$ in Figure 3-3 has a tension of 800 N. What is the moment about O of the force from this cable acting on the mast?

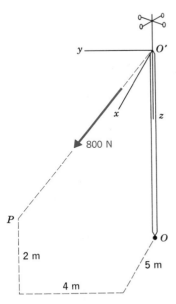

Since point P is on the line of action of this force, a position vector $\mathbf{r}$ from O to P can be used for evaluation of the moment about O. Referring to the axes shown in Figure 3-3, this vector has the resolution

$$\mathbf{r} = (5 \text{ m})\mathbf{u}_x + (4 \text{ m})\mathbf{u}_y + (-2 \text{ m})\mathbf{u}_z$$

The resolution of the force can be determined by multiplying its magnitude by the unit vector in the direction of $O'P$, components of which were evaluated in the previous example:

$$\mathbf{f} = 800 \text{ N } (0.4880 \ \mathbf{u}_x + 0.3904 \ \mathbf{u}_y + 0.7807 \ \mathbf{u}_x)$$
$$= (390.4 \text{ N})\mathbf{u}_x + (312.3 \text{ N})\mathbf{u}_y + (624.6 \text{ N})\mathbf{u}_z$$

The moment can now be evaluated with reference to Equation 3-22:

$$\mathbf{M}_O = [(4 \text{ m})(624.6 \text{ N}) - (-2 \text{ m})(312.3 \text{ N})]\mathbf{u}_x$$
$$+ [(-2 \text{ m})(390.4 \text{ N}) - (5 \text{ m})(624.6 \text{ N})]\mathbf{u}_y$$
$$+ [(5 \text{ m})(312.3 \text{ N}) - (4 \text{ m})(390.4 \text{ N})]\mathbf{u}_z$$
$$= (3123 \text{ N·m})\mathbf{u}_x - (3904 \text{ N·m})\mathbf{u}_y$$

Observe that point O' is also on the line of action of $\mathbf{f}$, so that a position vector $\mathbf{r} = (-10 \text{ m})\mathbf{u}_z$ could have been used instead of the one above, and with less arithmetic.

The *moment about an axis OP* is defined as the projection onto the axis of the moment about some point on the axis. This can be expressed analytically in terms of a unit vector $\mathbf{u}_{OP}$ along the axis:

$$\boxed{\mathbf{M}_{OP} = (\mathbf{M}_O \cdot \mathbf{u}_{OP})\mathbf{u}_{OP}} \tag{3-25a}$$

Useful insight can be gained by examining the following rearranged form:

$$\mathbf{M}_{OP} = [(\mathbf{r} \times \mathbf{f}) \cdot \mathbf{u}_{OP}]\mathbf{u}_{OP}$$
$$= [(\mathbf{u}_{OP} \times \mathbf{r}) \cdot \mathbf{f}]\mathbf{u}_{OP} \tag{3-25b}$$

Now resolve $\mathbf{f}$ into a component $\mathbf{f}_M$ perpendicular to the plane of $\mathbf{r}$ and OP, a component $\mathbf{f}_P$ parallel to OP, and a component $\mathbf{f}_N$ perpendicular to

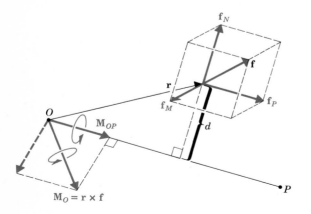

$\mathbf{f}_M$ and $\mathbf{f}_P$. Noting that $\mathbf{u}_{OP} \times \mathbf{r}$ is perpendicular to $\mathbf{f}_P$ and $\mathbf{f}_N$, we see from (3-25b) that these two force components do not contribute to $\mathbf{M}_{OP}$. Also, because $\mathbf{u}_{OP} \times \mathbf{r}$ has the magnitude $r \sin \measuredangle_{OP}^{\mathbf{r}}$, we can express the magnitude of the moment component as

$$\boxed{M_{OP} = f_M d}$$

(3-25c)

where d is the perpendicular distance from the head of $\mathbf{r}$ to the axis OP.

The sketch of the mast in the following example indicates how this idea may be used to compute moments about the axes through O and parallel to the x and y axes.

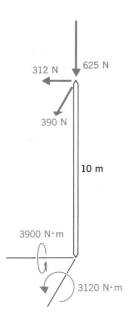

The *resultant moment about point O of a set of forces* $\mathbf{f}_1$, $\mathbf{f}_2$, $\mathbf{f}_3$, . . . is defined as the sum of the moments of the individual forces in the set:

$$\mathbf{M}_O = \mathbf{r}_1 \times \mathbf{f}_1 + \mathbf{r}_2 \times \mathbf{f}_2 + \mathbf{r}_3 \times \mathbf{f}_3 + \cdot \cdot \cdot$$

(3-26)

where $\mathbf{r}_i$ is a position vector from point O to any point on the line of action of $\mathbf{f}_i$ ($i = 1,2,3, \ldots$).

The *resultant moment about an axis* OP *of a set of forces* is defined as

$$\mathbf{M}_{OP} = [(\mathbf{r}_1 \times \mathbf{f}_1 + \mathbf{r}_2 \times \mathbf{f}_2 + \mathbf{r}_3 \times \mathbf{f}_3 + \cdot \cdot \cdot) \cdot \mathbf{u}_{OP}]\mathbf{u}_{OP}$$
$$= [(\mathbf{u}_{OP} \times \mathbf{r}_1) \cdot \mathbf{f}_1 + (\mathbf{u}_{OP} \times \mathbf{r}_2) \cdot \mathbf{f}_2 + \cdot \cdot \cdot]\mathbf{u}_{OP}$$

(3-27)

By comparing the individual terms of (3-27) with (3-25b), we see that the resultant moment about an axis can be computed as the sum of the moments of the individual forces about the axis.

Example

The forces T_A, T_B, R_x, and R_z are those imparted to the sign pole in Figure 3-4 by the supports. Evaluate the resultant moment of all the forces about the axis AB.

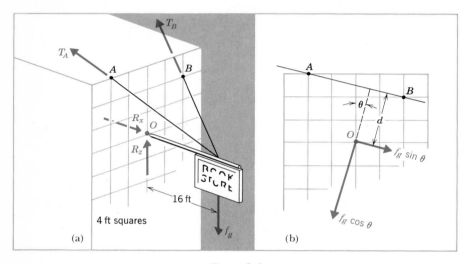

Figure 3-4

Since the lines of action of T_A, T_B, and R_z all pass through this axis, these forces will not contribute to the moment about AB. In Figure 3-4b we can see the components of the gravitational force in directions parallel and perpendicular to AB, and the moment arm d for the force R_x. The resultant moment about the axis AB is then

$$M_{AB} = f_g \cos \theta \, (16 \text{ ft}) - R_x \, d$$
$$= [(16 \text{ ft}) f_g - (10 \text{ ft}) R_x] \cos \theta$$

Varignon's Theorem. Suppose the lines of action of several forces intersect at some point P. Then the same position vector $\mathbf{r}$, from point O to point P, can be used to evaluate the moment about O of each of the forces, and the resultant moment about O of all the forces can be written as

$$\mathbf{M}_O = \mathbf{r} \times \mathbf{f}_1 + \mathbf{r} \times \mathbf{f}_2 + \mathbf{r} \times \mathbf{f}_3 + \cdots$$

But in view of Equation 3-11, this may also be written as

$$\mathbf{M}_O = \mathbf{r} \times (\mathbf{f}_1 + \mathbf{f}_2 + \mathbf{f}_3 + \cdots)$$

Therefore, *the sum of the moments of the force components is equal to the moment of the resultant force.* This theorem was given by Pierre Varignon in his book, *Nouvelle Méchanique ou Statique* (1725).

The usefulness of this result is illustrated for the special planar case in Section 2-3. The sketch on page 49 also indicates how the components of the resultant moment can be determined in terms of a set of components of the resultant force.

Equivalent Force Systems.   For the analysis of *rigid* bodies, two sets of forces are said to be equivalent if the sum of the forces in one set is equal to that of the other, and if, about any point, the resultant moment of the forces in one set is equal to that of the other.

To check equivalence, we must compare the sums of the force vectors and the moment resultant with respect to just one, arbitrarily selected point. For example, we checked the equivalence of the force system in Figure 2-8 in terms of the moments about the corner of the bracket and then observed that the moment about the free end was *also* the same for each set of forces. A further check of the moments about the point of attachment at the wall will show agreement again.

To demonstrate this in general, let us denote two different sets of forces as follows:

	SET 1	SET 2
Forces	$f_1, f_2, \ldots f_m$	$g_1, g_2, \ldots g_n$
Position vectors emanating from point O	$r_1, r_2, \ldots r_m$	$s_1, s_2, \ldots s_n$
Resultant force	$f = f_1 + f_2 + \cdots + f_m$	$g = g_1 + g_2 + \cdots + g_n$
Resultant moment about O	$M_O = r_1 \times f_1 + r_2 \times f_2 + \cdots + r_m \times f_m$	$N_O = s_1 \times g_1 + s_2 \times g_2 + \cdots + s_n \times g_n$

Now, if the stated requirements are met,

$$f = g$$
$$M_O = N_O$$

Now consider the moments about a different point, O', located by a position vector ρ from O' to O. The resultant moment about O' of the first set

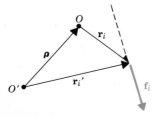

of forces is

$$
\begin{aligned}
\mathbf{M}_{O'} &= \mathbf{r}_1' \times \mathbf{f}_1 + \mathbf{r}_2' \times \mathbf{f}_2 + \cdots + \mathbf{r}_m' \times \mathbf{f}_m \\
&= (\mathbf{r}_1 + \boldsymbol{\rho}) \times \mathbf{f}_1 + (\mathbf{r}_2 + \boldsymbol{\rho}) \times \mathbf{f}_2 + \cdots + (\mathbf{r}_m + \boldsymbol{\rho}) \times \mathbf{f}_m \\
&= \mathbf{r}_1 \times \mathbf{f}_1 + \mathbf{r}_2 \times \mathbf{f}_2 + \cdots + \mathbf{r}_m \times \mathbf{f}_m + \boldsymbol{\rho} \times (\mathbf{f}_1 + \mathbf{f}_2 + \cdots + \mathbf{f}_m) \\
&= \mathbf{M}_O + \boldsymbol{\rho} \times \mathbf{f}
\end{aligned}
\tag{3-28}
$$

Similarly, the resultant moment about O' of the second-set of forces is

$$
\mathbf{N}_{O'} = \mathbf{N}_O + \boldsymbol{\rho} \times \mathbf{g}
$$

But since $\mathbf{N}_O = \mathbf{M}_O$ and $\mathbf{g} = \mathbf{f}$,

$$
\mathbf{N}_{O'} = \mathbf{M}_{O'}
$$

That is, *equality of force resultant and moment resultant about one point implies equality of moment resultant about every other point.*

Do not forget that equivalence in the sense that it is used here implies an equivalent response in a *rigid* body; the response in a *deformable* body depends on each individual point of application of the forces in the set.

Composition of Force Systems. The composition of a force system is the process of combining several components to form a simpler equivalent set. For example, let us consider the planar set of forces shown in Figure 3-5a.

A convenient way of determining the resultant force is to resolve each force into horizontal and vertical components and add, as follows:

$$
\begin{aligned}
\mathbf{f}_A &= (-6 \text{ N})\mathbf{u}_x - (2 \text{ N})\mathbf{u}_y \\
\mathbf{f}_B &= (20 \text{ N})\mathbf{u}_x - (15 \text{ N})\mathbf{u}_y \\
\mathbf{f}_C &= (10 \text{ N})\mathbf{u}_x + (24 \text{ N})\mathbf{u}_y \\
\hline
\mathbf{f} &= (24 \text{ N})\mathbf{u}_x + (7 \text{ N})\mathbf{u}_y
\end{aligned}
$$

Thus the replacement of the given forces with this resultant will satisfy the first requirement for forming an equivalent force system. To effect moment equivalence, let us first evaluate the moment resultant of the given system of forces about point O':

$$
\begin{aligned}
\mathbf{M}_{O'(A)} &= 20 \text{ N·m } \mathbf{u}_z \\
\mathbf{M}_{O'(B)} &= -35 \text{ N·m } \mathbf{u}_z \\
\mathbf{M}_{O'(C)} &= 90 \text{ N·m } \mathbf{u}_z \\
\hline
\mathbf{M}_{O'} &= 75 \text{ N·m } \mathbf{u}_z
\end{aligned}
$$

Now, an equivalent force system would be a single force $\mathbf{f} = (24 \text{ N})\mathbf{u}_x + (7 \text{ N})\mathbf{u}_y$ with its line of action passing through O' (contributing

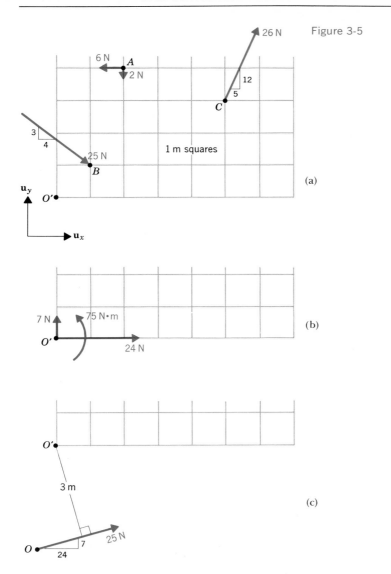

Figure 3-5

(a)

(b)

(c)

no moment about O'), together with a couple of moment $\mathbf{M}_{O'} = (75 \text{ N·m})\mathbf{u}_z$, as is depicted in Figure 3-5b. Further simplification can be achieved by replacing this system with a single force $\mathbf{f}$ with its line of action placed at a distance $\rho = (75 \text{ N·m})/25 \text{ N} = 3 \text{ m}$ from point O', as shown in Figure 3-5c.

By this process, any planar system of forces can be reduced to an equivalent single force with a specified line of action. However, the same is not true for a general three-dimensional system of forces; in general, the nearest we can come to achieving this is to eliminate the component of moment perpendicular to the force.

There can always be found a line of action for the force such that the accompanying moment is parallel to the force. This particular force system is called a *wrench*. Given a set of forces $\mathbf{f}_1, \mathbf{f}_2, \mathbf{f}_3, \cdots$ with specified lines of action, the equivalent wrench can be determined as follows.

1. The resultant force is evaluated by addition:

$$\mathbf{f} = \mathbf{f}_1 + \mathbf{f}_2 + \mathbf{f}_3 + \cdots$$

2. A point O' is selected and the resultant moment about this point is computed:

$$\mathbf{M}_{O'} = \mathbf{r}_1' \times \mathbf{f}_1 + \mathbf{r}_2' \times \mathbf{f}_2 + \mathbf{r}_3' \times \mathbf{f}_3 + \cdots$$

3. This moment vector is then resolved into a component parallel to $\mathbf{f}$ and a component perpendicular to $\mathbf{f}$ (Equation 3-15):

$$\mathbf{M}_{O'} = \left(\frac{\mathbf{f} \cdot \mathbf{M}_{O'}}{\mathbf{f} \cdot \mathbf{f}}\right) \mathbf{f} + \left(\frac{\mathbf{f} \times \mathbf{M}_{O'}}{\mathbf{f} \cdot \mathbf{f}}\right) \times \mathbf{f}$$

Comparing this equation with (3-28),

$$\mathbf{M}_{O'} = \mathbf{M}_O + \boldsymbol{\rho} \times \mathbf{f} \qquad\qquad [3\text{-}28]*$$

we note that the vector

$$\boldsymbol{\rho} = \frac{\mathbf{f} \times \mathbf{M}_{O'}}{\mathbf{f} \cdot \mathbf{f}} \qquad\qquad (3\text{-}29)$$

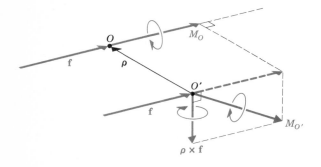

locates a point O relative to O' such that the parallel component $[(\mathbf{f} \cdot \mathbf{M}_{O'})\mathbf{f}]/\mathbf{f} \cdot \mathbf{f}$ is the resultant moment about O. The wrench therefore consists of a force $\mathbf{f}$ with line of action passing through point O,

---

* Brackets signify an equation that is repeated from an earlier development.

together with a couple having moment given by

$$\mathbf{M}_O = \frac{(\mathbf{f} \cdot \mathbf{M}_{O'})\mathbf{f}}{\mathbf{f}^2} \tag{3-30}$$

Example

Determine the wrench equivalent to the three forces shown in Figure 3-6a.

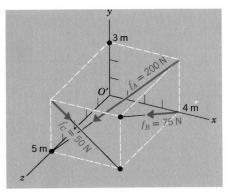

Figure 3-6

(a)

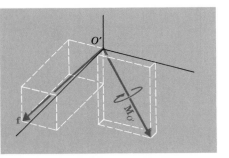

(b)

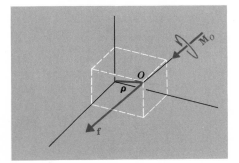

(c)

Resolutions of the given forces along the axes shown are

$$\mathbf{f}_A = (-113.137 \text{ N}) \ \mathbf{u}_x + (-84.853 \text{ N}) \ \mathbf{u}_y + (141.421 \text{ N}) \ \mathbf{u}_z$$
$$\mathbf{f}_B = \qquad\qquad\qquad (38.587 \text{ N}) \ \mathbf{u}_y + (64.312 \text{ N}) \ \mathbf{u}_z$$
$$\mathbf{f}_C = \quad (40.000 \text{ N}) \ \mathbf{u}_x + (-30.000 \text{ N}) \ \mathbf{u}_y$$

and position vectors locating points on the lines of action are

$$\mathbf{r}'_A = (5 \text{ m})\mathbf{u}_z$$
$$\mathbf{r}'_B = (4 \text{ m})\mathbf{u}_x$$
$$\mathbf{r}'_C = (3 \text{ m})\mathbf{u}_y + (5 \text{ m})\mathbf{u}_z$$

The resultant force is found by summation:

$$\mathbf{f} = (-73.137 \text{ N}) \ \mathbf{u}_x + (-76.266 \text{ N}) \ \mathbf{u}_y + (205.733 \text{ N}) \ \mathbf{u}_z$$

Next, the resultant moment about O' is

$$\mathbf{M}_{O'} = \mathbf{r}'_A \times \mathbf{f}_A + \mathbf{r}'_B \times \mathbf{f}_B + \mathbf{r}'_C \times \mathbf{f}_C$$
$$= (574.26 \text{ N·m})\mathbf{u}_x - (622.93 \text{ N·m})\mathbf{u}_y + (34.35 \text{ N·m})\mathbf{u}_z$$

An equivalent system would consist of the above resultant force $\mathbf{f}$ acting through O', plus a couple having moment equal to $\mathbf{M}_{O'}$ above, as is shown in Figure 3-6*b*. The wrench consists of the force $\mathbf{f}$ acting through the point O, which is located relative to O' by the position vector

$$\boldsymbol{\rho} = \frac{\mathbf{f} \times \mathbf{M}_{O'}}{\mathbf{f} \cdot \mathbf{f}}$$

$$= \frac{(125 \ 540 \text{ N}^2\text{·m}) \ \mathbf{u}_x + (120 \ 660 \text{ N}^2\text{·m}) \ \mathbf{u}_y + (89 \ 360 \text{ N}^2\text{·m}) \ \mathbf{u}_z}{53 \ 490 \text{ N}^2}$$

$$= (2.347 \text{ m}) \ \mathbf{u}_x + (2.256 \text{ m}) \ \mathbf{u}_y + (1.670 \text{ m}) \ \mathbf{u}_z$$

together with a couple having moment

$$\mathbf{M}_O = \frac{\mathbf{f} \cdot \mathbf{M}_{O'}}{\mathbf{f} \cdot \mathbf{f}} \ \mathbf{f}$$

$$= \frac{12 \ 575 \text{ N}^2\text{·m}}{53 \ 490 \text{ N}^2} \ [(-73.137 \text{ N}) \ \mathbf{u}_x - (76.266 \text{ N}) \ \mathbf{u}_y + (205.733 \text{ N}) \ \mathbf{u}_z)]$$

$$= (-17.19 \text{ N·m}) \ \mathbf{u}_x - (17.93 \text{ N·m}) \ \mathbf{u}_y + (48.36 \text{ N·m}) \ \mathbf{u}_z$$

This is illustrated in Figure 3-6*c*.

Problems

3-31   Evaluate the moments of the forces T_P and T_Q exerted on the mast in Figure 3-3 by the cables $O'P$ and $O'Q$, using a position vector from O to O'. Compare your result with values determined by using position vectors from O to P and O to Q.

3-32 Evaluate the moments about O of the forces T_Q and T_R exerted on the mast in Figure 3-3 by the cables $O'Q$ and $O'R$. If the resultant moment about O of the forces T_P (= 800 N), T_Q, and T_R is equal to **0**, what are the values of T_Q and T_R? What is the resultant of the three forces from the cables?

3-33 Evaluate the moment about the axis OP of the force T_Q exerted by the cable $O'Q$ on the mast in Figure 3-3. From this determine the perpendicular distance between the lines OP and $O'Q$.

3-34 Evaluate the moment about the axis OQ of the force T_P exerted by the cable $O'P$ on the mast in Figure 3-3. From this determine the perpendicular distance between the lines OQ and $O'P$.

3-35 Calculate the sum of the squares of the direction cosines for the line $O'P$ in Figure 3-3 (see page 45). Repeat the calculation for the line $O'Q$. Will the result be true in general?

3-36 In terms of the quantities shown on the sketch, evaluate f_x, f_y, and f_z. In terms of these values evaluate $f_x^2 + f_y^2 + f_z^2$.

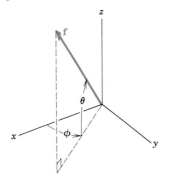

3-37 Show that the moment of a couple is independent of the point about which the moment is reckoned.

3-38 Determine the single force and its line of action, equivalent to the system in Figure 3-5a, by computing the resultant moment about point B rather than point O.

3-39 (a) Evaluate the equivalent force-couple system acting at the origin.

(b) Evaluate the equivalent wrench.

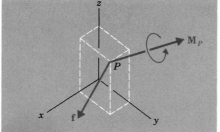

Point P: (2 ft, 3 ft, 6 ft)

$$f = 13 \text{ lb}; f_x = 12 \text{ lb}; f_y = 5 \text{ lb}.$$

$$\mathbf{M}_P = -18 \text{ lb·ft } \mathbf{u}_x + 9 \text{ lb·ft } \mathbf{u}_y$$
$$+ 6 \text{ lb·ft } \mathbf{u}_z$$

3-40   During a wind gust, the airplane feels aerodynamic loads equivalent to the four forces shown. Points P_1, P_2, and P_3 are each 5 m from the origin, and P_3 is 1.5 m from the y axis. Determine the equivalent wrench.

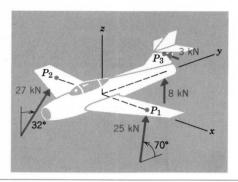

STATIC EQUILIBRIUM

Problems in statics entail the determination of unknown forces that are induced by known forces acting on bodies at rest. For the solution of these problems there are two fundamental laws: that of force equilibrium and that of moment equilibrium.

Force equilibrium is the law that the sum of all forces acting on a body at rest is zero:

$$\boxed{\mathbf{f} = \mathbf{0}} \qquad (4\text{-}1a)$$

This law is a special case of Newton's second law of motion, $\mathbf{f} = m\mathbf{a}$.

Moment equilibrium is the law that *the resultant moment about any point of all forces acting on a body at rest is zero:*

$$\boxed{\mathbf{M} = \mathbf{0}} \qquad (4\text{-}2a)$$

This law is a special case of a law of rotational motion, which we will derive from Newton's second and third laws.

Each of the laws, (4-1a) and (4-2a), is a vector equation in three dimensions, implying equality of components in three different directions.

Resolving all forces into a set of mutually perpendicular directions usually provides the best basis for problem solution. After these directions are selected, the equation of force equilibrium may be written in the form

$$(\Sigma f_x)\mathbf{u}_x + (\Sigma f_y)\mathbf{u}_y + (\Sigma f_z)\mathbf{u}_z = \mathbf{0}$$

in which Σf_x, Σf_y, and Σf_z are the summations of force components in the three directions. This vector equation implies the three component equations

$$
\begin{aligned}
\Sigma f_x &= 0 \\
\Sigma f_y &= 0 \\
\Sigma f_z &= 0
\end{aligned}
\qquad (4\text{-}1\mathrm{b})
$$

Similar resolution of the moment resultant about a selected point O leads to the three equations

$$
\begin{aligned}
\Sigma M_{Ox} &= 0 \\
\Sigma M_{Oy} &= 0 \\
\Sigma M_{Oz} &= 0
\end{aligned}
\qquad (4\text{-}2\mathrm{b})
$$

in which ΣM_{Ox}, ΣM_{Oy}, and ΣM_{Oz} are the summations of components of moments about O in the three directions.

In general, then, we can expect to have *six* independent equations of equilibrium for each body considered.

In cases in which all the forces act in one plane, one of the three equations from (4-1) becomes trivial ($0 = 0$). Also, the resultant moment of the forces about any point in the plane will be perpendicular to this plane, so that (4-2) yields only one nontrivial equation. Thus, in planar systems, we can expect to have *three* equations of equilibrium for each body.

4–1

FREE BODY DIAGRAMS. To apply (4-1) and (4-2) it is necessary to identify carefully a specific body (which may be a part of a larger structural system) and consider *all* the forces acting on the body. Experience has shown that to consistently do this correctly, it is necessary to draw a *free-body diagram.*

To illustrate this procedure, let us consider the device shown in Figure 4-1. A number of different free-bodies are possibly useful, so we will construct several of them.

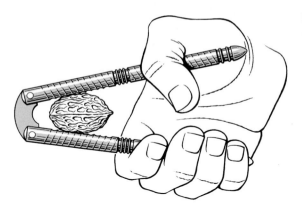

Figure 4-1

What To Exclude and Include. First, consider the system consisting of the nutcracker together with the walnut. This system is shown isolated from all other objects, with arrows depicting the forces that come from objects *external* to the system. Assuming forces of gravity to be negligible, the only external body that exerts forces on this system is the hand. Therefore, the appropriate arrows are those shown in Figure 4-2a. The

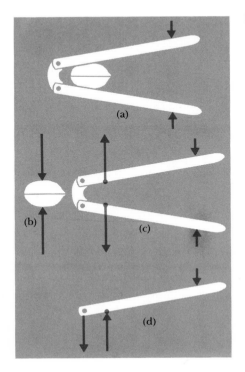

Figure 4-2

(a)

(b)

(c)

(d)

interaction between the nut and the cracker is *not* shown on this free-body, since this is internal to this system.

To expose the force tending to break the nut, we might consider free-bodies of the nut and of the nutcracker (Figure 4-2*b* and *c*). Now, external to the nutcracker are the hand and the walnut. Therefore, arrows depicting the forces from the nut as well as from the hand are included on the free-body of the nutcracker.

Another possibly useful free-body would be the upper handle. Objects external to this are the hand, the walnut, and the connecting pin. This free-body appears in Figure 4-2*d*.

Labeling.   To show clearly the physical significance of quantities in written equations, magnitudes of forces must be properly identified on the free-body diagrams. The implication of the letter placed beside an arrow for this purpose must be precisely understood. The letter represents a *scalar* multiplier of a unit vector in the direction of the arrow. Thus, since this quantity can take on positive or negative values, a force in the same or opposite direction as indicated by the arrow can be represented.

Figure 4-3 indicates appropriate labeling of our free-bodies for analy-

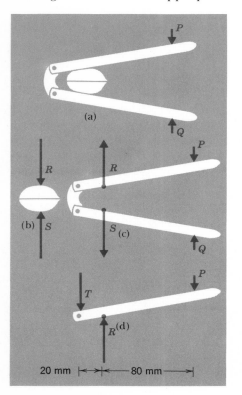

Figure 4-3

sis of the nut-cracking system. If an analysis leads to the values, say, $P = 80$ N and $R = 400$ N, this would imply that the forces acting on the upper handle are 80 newtons *downward* on the right-hand end and 400 newtons *upward* where the walnut makes contact. If different values led to, say, $P = -15$ N and $R = -75$ N, these values would imply that the forces acting on the upper handle are 15 newtons *upward* on the right-hand end and 75 newtons *downward* where the walnut makes contact. (A little epoxy between the walnut and the nutcracker would make this possible.)

Observe that the forces on the walnut have the same labels as their counterparts on the nutcracker, and the arrows have opposite directions. We have in this way implied satisfaction of Newton's third law without further fuss. Other relatively simple aspects of force analysis can be taken care of as the free-bodies are constructed; for example, unless the walnut is to accelerate, it is evident from a glance at its free-body that $R = S$. Writing this equation can be circumvented by simply labeling both arrows with the same letter. Can you find another free-body in Figure 4-3 in which some analysis can be taken care of as the labeling is done?

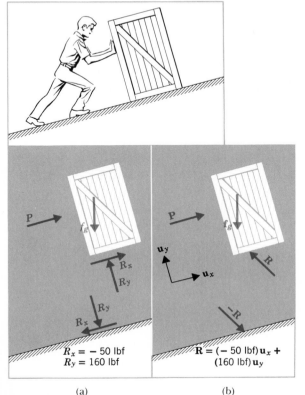

Figure 4-4

$R_x = -50$ lbf
$R_y = 160$ lbf

$\mathbf{R} = (-50 \text{ lbf})\mathbf{u}_x + (160 \text{ lbf})\mathbf{u}_y$

(a)   (b)

Occasionally, a force depicted on a free-body diagram is labeled with a *boldface* letter. This, of course, implies a *vector* value, which can take on any direction and magnitude. In this case the labeling appears differently, as a comparison of (a) and (b) in Figure 4-4 shows. Notice the necessity of indicating the base directions with $\mathbf{u}_x$ and $\mathbf{u}_y$ before a specific value can be identified with the boldface labeling. Also, note that the consistent manner of labeling the oppositely-directed reaction $-\mathbf{R}$ incorporates a minus sign, whereas in Figure 4-4a the force components acting on the incline have positive signs on their labels.

Problems

4-1   The sketches depict a number of the types of connections between structural members. On a partial* free-body of the member like that indicated on the right, show the components of force and moment which the connection is capable of transmitting.

(a)

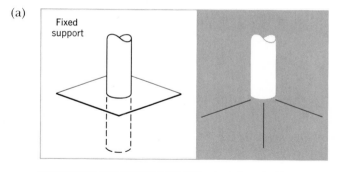

(b)

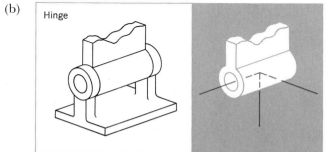

* A complete free body would include the force transmitted from the portion of the member removed; ignore this.

(c)

Sleeve bearing

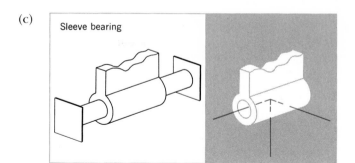

(d)

Spline

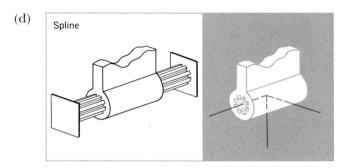

(e)

Ball
and
socket

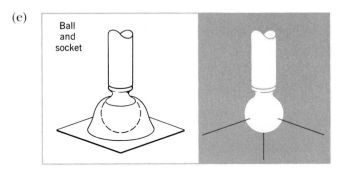

(f)

Smooth (slippery)
surface

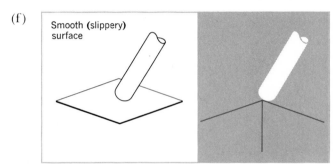

(g)

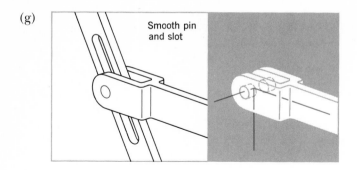

Smooth pin and slot

4-2   Draw a free-body diagram of:
  (a) The pulley.
  (b) The frame.
  (c) The system consisting of the frame, pulley, line, block, and man.

4-3   Draw a free-body diagram of:
  (a) Member A.
  (b) Member B.
  (c) Member C.
  (d) The block M.
  (e) The system consisting of A, B, C, and M.

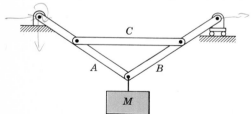

4-4   Draw a free-body diagram of:
(a) The rod *A*.   (b) The lifeboat davit *B*.
(c) The lifeboat *C*.   (d) The system consisting
                                                      of *A*, *B*, and *C*.

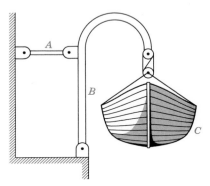

4-5   Draw a free-body diagram of the flagpole and flag.

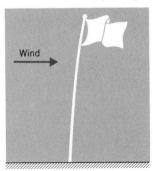

4-6   Draw a free-body diagram of each of the logs in the pile.

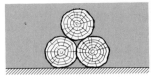

4-7   Draw a free-body diagram of the hand of Figure 1-1.

4-8   Draw a free-body diagram of each of the two large parts of the
channel-lock pliers.

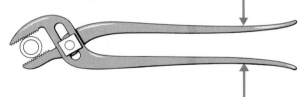

$$4-2$$

EQUATIONS OF EQUILIBRIUM.   With properly labeled free-body diagrams drawn, equations of equilibrium can be written for any of the chosen bodies. For example, considering the vertical direction for force equilibrium of the free-body of Figure 4-3a, we can write

$$Q - P = 0$$

Each of the other two equations of force equilibrium, involving components in the horizontal direction and components perpendicular to the page, is the trivial $0 = 0$. Account of moment equilibrium tells us that the lines of action of the two forces must coincide, a fact that has already been incorporated in the free-body diagram. Similar analysis of the free-bodies of Figure 4-3b and c leads to

$$S - R = 0$$
$$Q - P + R - S = 0$$

Now suppose that the reason for this analysis is to obtain an estimate of how hard one must squeeze in order to crack a nut, given that cracking requires 645 N applied to the nut. Then

$$R = 645 \text{ N}$$

and the three equations above contain the three unknowns P, Q, and S. Unfortunately, attempts to solve these for P or Q will fail, because the three equations are not independent. (The last equation can be deduced from the first two by addition.) Equilibrium of still another body will then have to be considered to obtain an independent equation. The free-body of the handle in Figure 4-3d can provide two more equations, the vertical component of force equilibrium,

$$T + P = 645 \text{ N} \tag{a}$$

and an equation of moment equilibrium. Summing moments about the point of contact with the walnut leads to

$$(20 \text{ mm})T - (80 \text{ mm})P = 0 \tag{b}$$

To solve for P, we can multiply equation (a) by 20 mm and subtract equation (b) from the result, yielding

$$(100 \text{ mm})P = (20 \text{ mm})(645 \text{ N}) \tag{c}$$

This gives

$$P = 129 \text{ N}$$

A more direct analysis stems from considering moments about a different point on the handle, resulting in (c) as the first equation written.

The following examples provide further illustration of the use of the basic considerations of static equilibrium.

Example

Determine the forces in the cable and in the boom shown in Figure 4-5.

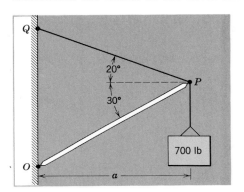

Figure 4-5

Free-body diagrams of the boom and of the suspended block are shown in Figure 4-6. First, equilibrium of the block requires that

$$T_1 = 700 \text{ lbf}$$

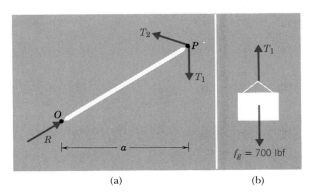

Figure 4-6

(a)   (b)

Considering moment equilibrium about point O of the boom, we can write

$$(T_2 \sin 20° - T_1)a + (T_2 \cos 20°)(a \tan 30°) = 0$$

from which

$$T_2 = \frac{T_1}{\sin 20° + \cos 20° \tan 30°}$$

$$= 791 \text{ lbf}$$

Considering moment equilibrium about point P of the boom, we see that the reaction R must be directed along the boom. Equilibrium of horizontal forces then gives

$$R \cos 30° - T_2 \cos 20° = 0$$

from which

$$R = \frac{T_2 \cos 20°}{\cos 30°}$$
$$= 859 \text{ lbf}$$

Example

Determine the tension in the cable AB holding up the crane boom in Figure 4-7.

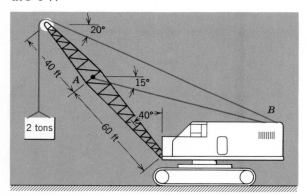

Figure 4-7

Free-body diagrams of the load being lifted, the pulley, and the boom are shown in Figure 4-8. Vertical force equilibrium of the free-body (a)

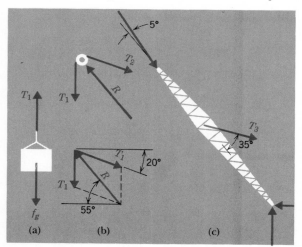

Figure 4-8

requires that
$$T_1 - f_g = 0$$

and moment equilibrium (about the pulley axle) of the free-body (b) requires that
$$T_1 r - T_2 r = 0$$

where r is the radius of the pulley. Incorporating this last result into the force diagram for the free-body (b), we see that the reaction R from the boom must be inclined at 55° from the horizontal. Horizontal force equilibrium for the pulley then requires that
$$T_2 \cos 20° - R \cos 55° = 0$$

Turning next to the free-body of the boom, we observe that the equation of moment equilibrium about point O will not contain the unknown reactions there; for this reason we write
$$R(100 \text{ ft} \sin 5°) - T_3(60 \text{ ft} \sin 35°) = 0$$

The above four equations contain the unknowns T_1, T_2, T_3, and R, and can be solved readily for these in terms of the force of gravity f_g:
$$T_1 = T_2 = f_g = 2 \text{ tons}$$
$$R = \frac{T_2 \cos 20°}{\cos 55°}$$
$$= 3.28 \text{ tons}$$
$$T_3 = \frac{100 \sin 5°}{60 \sin 35°} R$$
$$= 0.83 \text{ tons}$$

Note that the force of gravity on the boom has been neglected, so that the actual tension will be somewhat higher.

Example

Evaluate all the forces acting on each of the three members in the A-frame in Figure 4-9.

Figure 4-9

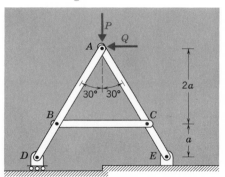

Figure 4-10 shows free-body diagrams of the overall structure and of the individual members. The roller support at D means that no horizontal

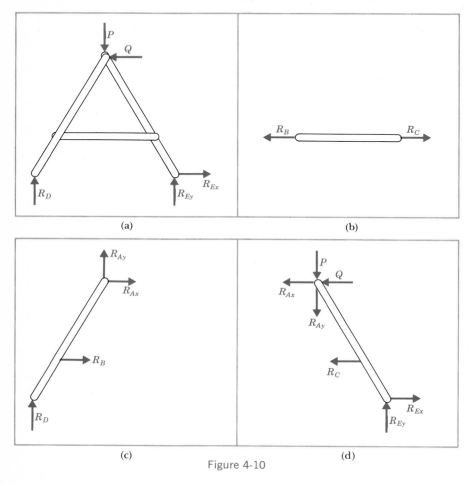

(a) (b)

(c) (d)

Figure 4-10

force can be transmitted to the ground at that point. From the free-body (a) we can write equations of equilibrium of moments about point E:

$$Q(3a) + P(3a \tan 30°) - R_D(6a \tan 30°) = 0$$

horizontal force equilibrium:

$$R_{Ex} - Q = 0$$

and equilibrium of moments about point D:

$$R_{Ey}(6a \tan 30°) + Q\,(3a) - P(3a \tan 30°) = 0$$

These equations can then be solved for the support reactions:

$$R_D = \frac{1}{2} P + \frac{\text{ctn } 30°}{2} Q$$

$$R_{Ex} = Q$$

$$R_{Ey} = \frac{1}{2} P - \frac{\text{ctn } 30°}{2} Q$$

As a check, you might examine vertical force equilibrium.

Moment equilibrium of the free-body (b) implies that the lines of action of R_B and R_C are along the bar. Its horizontal force equilibrium gives us

$$R_C - R_B = 0$$

Now turning to the free-body (c), we can write equations of moment equilibrium about point A:

$$R_B(2a) - R_D(3a \tan 30°) = 0$$

and horizontal and vertical force equilibrium:

$$R_{Ax} + R_B = 0$$

$$R_{Ay} + R_D = 0$$

With values of R_D, R_{Ex}, and R_{Ey} above, these equations give the following values of the remaining unknown reactions.

$$R_B = R_C = \frac{3 \tan 30°}{4} P + \frac{3}{4} Q$$

$$R_{Ax} = -\frac{3 \tan 30°}{4} P - \frac{3}{4} Q$$

$$R_{Ay} = -\frac{1}{2} P - \frac{\text{ctn } 30°}{2} Q$$

Example

Determine the forces in the boom OC and the cables CA and CB, in terms of the gravitational force f_g, in Figure 4-11.

We need to resolve the forces into the $\mathbf{u}_x$, $\mathbf{u}_y$, and $\mathbf{u}_z$ directions, and for this we must determine the direction cosines of OC, CA, and CB. A unit vector in the direction of OC has the resolution

$$\mathbf{u}_{OC} = -\frac{\sqrt{(7 \text{ m})^2 - (3 \text{ m})^2}}{7 \text{ m}} \mathbf{u}_x + \frac{3 \text{ m}}{7 \text{ m}} \mathbf{u}_z$$

$$= -0.9035 \mathbf{u}_x + 0.4286 \mathbf{u}_z$$

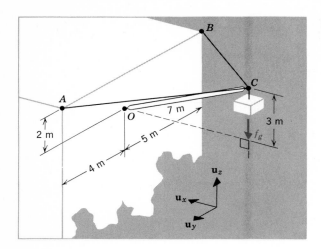

Figure 4-11

The lengths of the cables are

$$\overline{CA} = \sqrt{(7\text{ m})^2 - (3\text{ m})^2 + (4\text{ m})^2 + (1\text{ m})^2} = 7.5498\text{ m}$$
$$\overline{CB} = \sqrt{(7\text{ m})^2 - (3\text{ m})^2 + (5\text{ m})^2 + (1\text{ m})^2} = 8.1240\text{ m}$$

Unit vectors in the directions of the cables then have the resolutions

$$\mathbf{u}_{CA} = \frac{\sqrt{(7\text{ m})^2 - (3\text{ m})^2}}{7.550\text{ m}}\,\mathbf{u}_x + \frac{4\text{ m}}{7.550\text{ m}}\,\mathbf{u}_y - \frac{1\text{ m}}{7.550\text{ m}}\,\mathbf{u}_z$$
$$= 0.8377\,\mathbf{u}_x + 0.5298\,\mathbf{u}_y - 0.1325\,\mathbf{u}_z$$
$$\mathbf{u}_{CB} = \frac{\sqrt{(7\text{ m})^2 - (3\text{ m})^2}}{8.124\text{ m}}\,\mathbf{u}_x - \frac{5\text{ m}}{8.124\text{ m}}\,\mathbf{u}_y - \frac{1\text{ m}}{8.124\text{ m}}\,\mathbf{u}_z$$
$$= 0.7785\,\mathbf{u}_x - 0.6155\,\mathbf{u}_y - 0.1231\,\mathbf{u}_z$$

Now the forces shown in Figure 4-12 have the resolutions

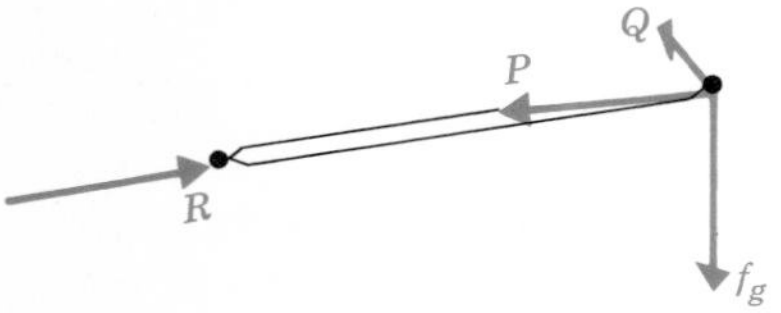

Figure 4-12

$$\mathbf{P} = P(0.8377\,\mathbf{u}_x + 0.5298\,\mathbf{u}_y - 0.1325\,\mathbf{u}_z)$$
$$\mathbf{Q} = Q(0.7785\,\mathbf{u}_x - 0.6155\,\mathbf{u}_y - 0.1231\,\mathbf{u}_z)$$
$$\mathbf{R} = R(-0.9035\,\mathbf{u}_x + 0.4286\,\mathbf{u}_y)$$
$$\mathbf{f}_g = -f_g\,\mathbf{u}_z$$

Let us now consider the moment equilibrium of the boom, about point O.

Each of the forces passes through point C, so a position vector

$$\mathbf{r} = (7 \text{ m}) \mathbf{u}_{OC}$$
$$= -6.3246 \text{ m } \mathbf{u}_x + 3.0000 \text{ m } \mathbf{u}_z$$

can be used for the evaluation of the moment of each force. The moment equation is

$$\mathbf{M}_O = \mathbf{r} \times \mathbf{P} + \mathbf{r} \times \mathbf{Q} + \mathbf{r} \times \mathbf{f}_g = \mathbf{0}$$

Or,

$$(-6.3246 \mathbf{u}_x + 3.0000 \mathbf{u}_z)$$
$$\times [(0.8377 \mathbf{u}_x + 0.5298 \mathbf{u}_y - 0.1325 \mathbf{u}_z)P$$
$$+ (0.7785 \mathbf{u}_x - 0.6155 \mathbf{u}_y - 0.1231 \mathbf{u}_z)Q - \mathbf{u}_z f_g] = \mathbf{0}$$
$$(-1.5894 \ P + 1.8464Q) \mathbf{u}_x$$
$$+ (1.6754P + 1.5570Q - 6.3246 \ f_g) \mathbf{u}_y$$
$$+ (-3.3508P + 3.8925Q) \mathbf{u}_z = \mathbf{0}$$

This vector equation implies three component equations, the first two of which are

$$-1.5894P + 1.8464Q = 0$$
$$1.6754P + 1.5570Q = 6.3246f_g$$

The third equation can be obtained from the first, and thus provides only a check. The above two equations can be solved simultaneously for the cable tensions:

$$P = 2.097f_g$$
$$Q = 1.805f_g$$

The compressive force in the boom can be determined by considering force equilibrium in any direction. It might be wise to sum forces in two or three directions as a check, since a computation as lengthy as the one above has many opportunities for errors.

$$R = 3.500f_g$$

Example

The cables OA, OB, and OC support the suspended block shown in Figure 4-13. Determine the tension in each cable in terms of the gravitational force f_g.

The desired tensions, P, Q, and R, are shown on the free body of the portion of the structure in the neighborhood of point O. To resolve the forces in the directions of $\mathbf{u}_x$, $\mathbf{u}_y$, and $\mathbf{u}_z$, the direction cosines between OA, OB, and OC and these directions must be evaluated. This is done, as

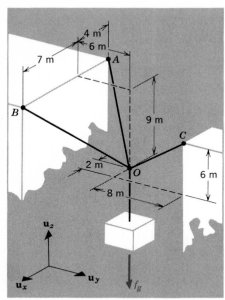

Figure 4-13

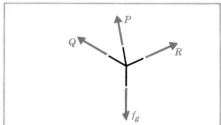

in the preceding example, by dividing the projection of the cable onto the axis by the length of the cable; for instance, the direction cosine between OA and $\mathbf{u}_x$ is

$$\cos \sphericalangle^{\mathbf{u}_x}_{OA} = \frac{-4 \text{ m}}{\sqrt{(4 \text{ m})^2 + (6 \text{ m})^2 + (9 \text{ m})^2}} = -0.346\,84$$

The results are summarized in the following:

$$\mathbf{u}_{OA} = -0.346\,84\ \mathbf{u}_x - 0.520\,27\ \mathbf{u}_y + 0.780\,40\ \mathbf{u}_z$$

$$\mathbf{u}_{OB} = 0.543\,31\ \mathbf{u}_x - 0.465\,69\ \mathbf{u}_y + 0.698\,54\ \mathbf{u}_z$$

$$\mathbf{u}_{OC} = 0.196\,12\ \mathbf{u}_x + 0.784\,46\ \mathbf{u}_y + 0.588\,35\ \mathbf{u}_z$$

Force equilibrium requires that

$$P\ \mathbf{u}_{OA} + Q\ \mathbf{u}_{OB} + R\ \mathbf{u}_{OC} - f_g\ \mathbf{u}_z = 0$$

With the above resolutions inserted, this vector equation implies the three component equations:

$$\Sigma f_x = 0.346\ 84\ P + 0.543\ 31\ Q + 0.196\ 12\ R = 0$$
$$\Sigma f_y = -0.520\ 27\ P - 0.465\ 69\ Q + 0.784\ 46\ R = 0$$
$$\Sigma f_z = 0.780\ 40\ P + 0.698\ 54\ Q + 0.588\ 35\ R = f_g$$

These three equations can be solved simultaneously for the three unknown forces. The results are

$$P = 0.6601 f_g$$
$$Q = 0.2169 f_g$$
$$R = 0.5666 f_g$$

Note the steps, each carefully taken, that are common to all the above problem solutions: free-body diagrams are drawn, known and unknown forces are identified and labeled, and equations of equilibrium for each free-body are written. These steps reduce the problem to one of computation, such as the solution of simultaneous algebraic equations.

The ability readily to apply these steps to new systems comes from *practicing* them; little further can be gained at this point by following additional worked-out examples. As this exercise is carried out the following should be kept in mind.

Many different equivalent sets of equations are possible, depending on which sets of objects are isolated as free-bodies, which directions are selected for vector resolutions, and which points are selected to compute moments. Therefore, if an analysis leads to a large set of equations with many unknowns, it is worth considering a different free-body, a different component resolution, or a different point for moment summation. For example, we could write three equations of equilibrium for the free-body in Figure 4-8c, containing the unknown force T_3 and the two reaction components at O. However, selection of point O for moment summation results in an equation that does not contain these last two unknown components, so that T_3 is directly solvable without recourse to additional equations.

Problems

4-9   The tension in cable $O'R$ in Figure 3-3 is equal to 700 lb. Determine the tensions in the cables $O'P$ and $O'Q$.

4-10 Evaluate the reaction at the wall induced by the 240-lb plumber of Problem 2-31.

4-11   Show why the tension on either side of the pulley is the same.

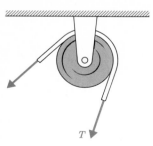

4-12   Determine the force with which the 80-kg man must pull on the rope in order to support himself.

4-13   Show that the moment M of the couple acting on the crank must equal Pb in order that the system be in static equilibrium.

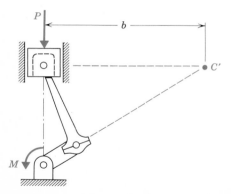

4-14 Each of the tracks in the upper pulley unit is recessed to fit the chain, so as to prevent slipping. The smaller track has a radius equal to 0.9 times that of the larger track. Estimate the force P necessary to lift the block by means of the differential chain hoist.

0.8 Mg

P

4-15 Determine the tension in the cable and the reaction at A.

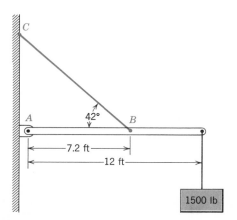

C

A

42°

B

7.2 ft

12 ft

1500 lb

4-16 Determine the force at each pin connection.

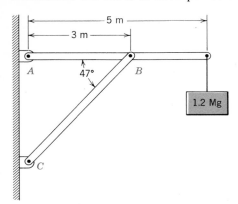

4-17 Determine the reactions at A, B, and O, induced by the 700-lb life-boat.

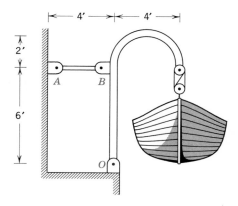

4-18 Evaluate the reactions at the supports for each case.

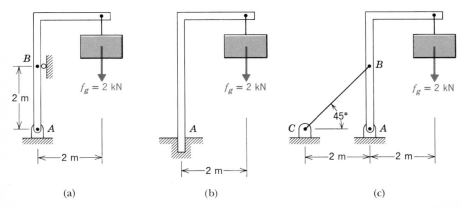

(a)   (b)   (c)

4-19   Determine the tension in the cable AB and the reaction at C.

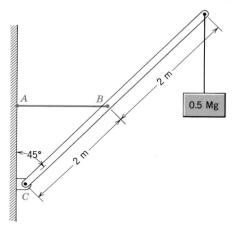

4-20   Determine the tension force T in terms of the gravitational force f_g.

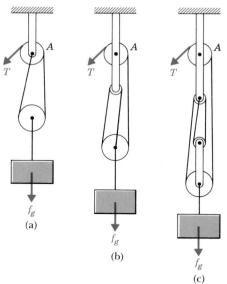

4-21   The surfaces are smooth where the drum makes contact. Determine the reactions at these points.

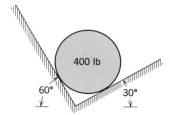

4-22 The surfaces of the bars are smooth where the drum makes contact. Determine the reactions at *A*, *B*, and *C*.

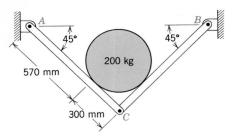

4-23 The 30-kg trapdoor is lifted by the corner as the illustration shows. Evaluate the reactions at the hinges at *A* and *B*.

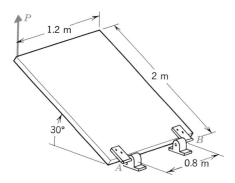

4-24 The gravitational forces acting on the cart and its load of coal are represented by the 600-lb force acting as shown. Determine the tension *T* required to hold the cart, and the reactions at the wheels.

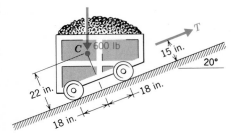

4-25 The piston slides with negligible friction inside a cylinder (not shown). Determine the reactions at the crank bearings, and the moment *M* of the couple acting on the crankshaft for equilibrium.

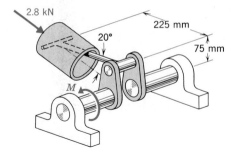

4-26 The maximum load that the dockyard crane must handle is $f_g = 20\,000$ lbf. The gravitational forces acting on the boom itself are represented by the 800-lb force at mid-span. Evaluate the tension in the cable and the reaction at O, and show how these vary with x.

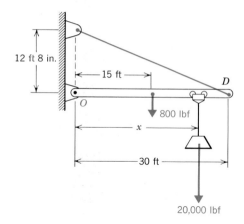

4-27 Evaluate the reactions at the trailer hitch and at each wheel.

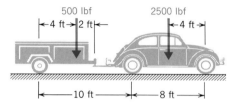

4-28 Two wheels, each of radius a but of different mass, are connected by the rod of length R. The assembly is free to roll in the circular trough. Determine the angle θ for equilibrium.

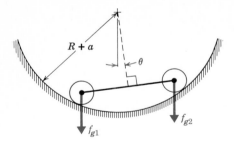

4-29 Determine the reaction at each bearing and the moment of the couple transmitted by the shaft at section A.

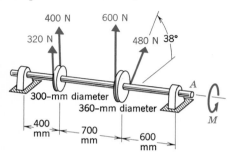

4-30 In the trailer "load-leveler" hitch, the angle bar slips into the cylindrical socket at A, forming a thrust bearing where it bottoms. The end B is then attached by a short chain to the towing vehicle. The pretension in the chain is 1.7 kN. Evaluate the reactions at A, C, D, and E.

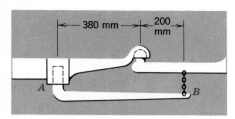

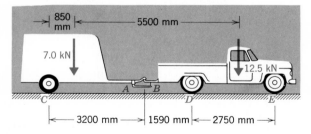

4-31 Evaluate all reactions between bars and at the supports.

(a)

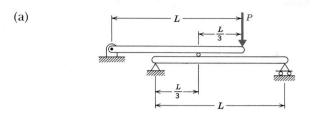

(b)

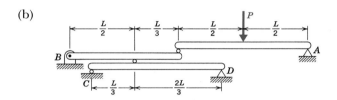

4-32 Evaluate the tensions in the guy lines AP and BQ, supporting the mast.

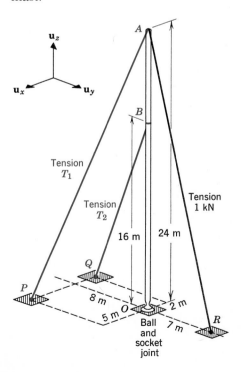

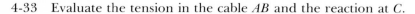

4-33 Evaluate the tension in the cable *AB* and the reaction at *C*.

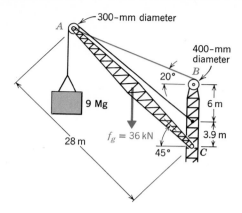

4-34 Compare the response of the system to the application of the couple at each of the two places. Analyze the situation carefully in terms of basic laws of equilibrium.

4-35 Determine the tensions in the cables *AB* and *CD*.

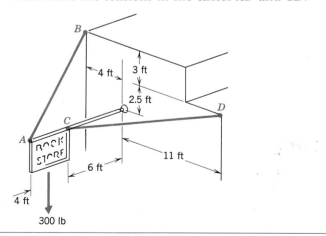

STATICALLY INDETERMINATE AND IMPROPER SUPPORT SYSTEMS.

Each of the bodies considered in the preceding examples is connected with its surroundings in such a way that it is possible to deduce the forces at these connections from equations of equilibrium.

Redundant Supports. If the degree of support is increased, this will generally increase the number of possible reaction components above the number of independent statics equations that can be written. The support system is then said to be *statically indeterminate*. For example, if the roller support at D in Figure 4-9 were replaced by the pinned type shown at E, two reaction components would be possible there in addition to the two at E. Since planar equilibrium of the free-body in Figure 4-10a can yield only three independent equations, the four reaction components could not be determined from statics alone.

To determine the forces acting on a body supported by a statically indeterminate system, *it is necessary to abandon the rigid body idealization and account for deformations of the body.* Such deformations are a primary focus of studies in the mechanics of deformable bodies (or "strength of materials") and will not be considered here.

Deficient Supports. If the support system is incapable of preventing motion when certain loads are applied, it is said to be *improper*. For example, if the support at E in Figure 4-9 were replaced by a roller type as at D, supports would be incapable of preventing horizontal motion.

Criteria. Usually the best procedure for determining whether a support system is indeterminate or improper or both is to consider the following tests. (a)   The support system is statically indeterminate if some degree of support can be removed, leaving the body still capable of withstanding any applied loads; or, equivalently, if reactions from the supports can exist in the absence of other applied forces. (b)   If the support system is incapable of preventing motion induced by all applied forces, additional support must be added to make the system proper.

The answers to these tests are nearly always obvious after simple inspection. However, there are situations where a more formal procedure may be needed. It can be shown that necessary and sufficient conditions for a proper support system are that it be capable of transmitting six forces with lines of action that cannot be intersected by a straight line.* In the planar case, the support system must be capable of transmitting three

---

* G. W. Housner and D. E. Hudson, *Applied Mechanics Statics*, Second Edition, D. Van Nostrand Co., New York, 1961.

forces with lines of action that do not intersect. In applying these criteria, parallel lines are considered to intersect at "infinity." Any supports in addition to those that meet these requirements are redundant, and their presence means that the support system is statically indeterminate.

Clamped Supports.   Certain types of attachements are idealized as transmitting locally applied couples and are called *clamped supports*. The above

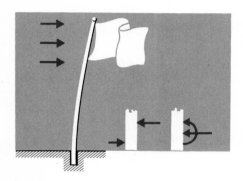

criteria can be applied to systems with such supports by considering the two-force equivalent of the force-couple combination at the clamp.

Problems

4-36   Is the mast and guy cable system shown in Figure 3-3 statically determinate?

4-37   Is the system of Problem 4-27 statically determinate? Is that of Problem 4-30 statically determinate?

4-38   Is the system of Problem 4-32 statically determinate?

4-39   Which of the following support systems are statically indeterminate, and which are improper?

(a)

(b)

(c)

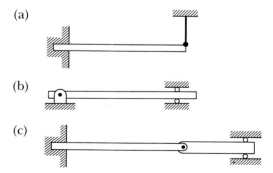

(d)

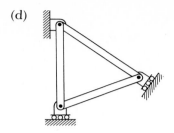

(e)   Rough
      surfaces

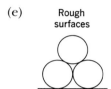

SOME SPECIAL APPLICATIONS

Most mechanical devices, structures, and machines have features that lead naturally to classification of some kind. To group them accordingly as our study proceeds is appropriate, because many of these special features are worthy of concentrated attention. However, important as they are, the special topics discussed in this and the next chapter are secondary to the basic equilibrium principles. Therefore, the most important reason for considering them now is to provide a further reinforcement of skills for drawing and labeling of free-body diagrams and for writing and solving equations of equilibrium.

5-1

TRUSSES. Trusslike structures are used to support bridge decks, roofs, power lines, and many other loads.

A truss is a structure that is built up with interconnected *axial force members*. Such a member is a straight rod that can transmit force along its axis but it does not carry a couple. The limitation to axial force results from the fact that the interconnections within a truss are of the ball and

socket type; that is, they constrain the end points of the connected members against relative position change, but allow the members complete freedom to rotate about the connection point. Also, external forces are applied only at these joints.

Each member, then, is subjected to only two forces, one at each end. From moment equilibrium it follows that the lines of action of the two forces must coincide with the line connecting the points of application.

Real trusslike structures normally have joints that can actually transmit couples. However, when the members are relatively long and slender,

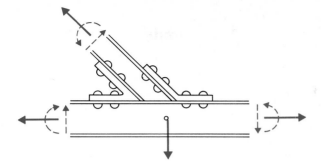

and loads are applied only near the joints, the idealization as a truss has been found to lead to very good estimates of the internal forces. And because the idealization leads to significant simplification of analysis, it is well justified.

Notation. In analyzing trusses, it is convenient to adopt a uniform notation for specifying the forces within the structure. Here we specify the various joints with letters, A, B, C, and so forth, and the *tensile* force in the

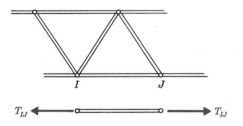

member connecting the joints I and J as T_{IJ} ($I, J = A, B, C, \cdot \cdot \cdot$). Thus if the value of T_{IJ} is positive, the bar is in tension, whereas if the value of T_{IJ} is negative, the bar is in compression. This convention must be kept in mind as free-body diagrams are labeled, and as results are interpreted.

Equations from Joints. One approach to determining the forces in the individual members within a truss is to isolate as a free-body the portion of the structure in the neighborhood of each joint, and to write equations of force equilibrium for each of them. This procedure is illustrated in terms of the plane truss depicted in Figure 5-1. The free-body diagrams

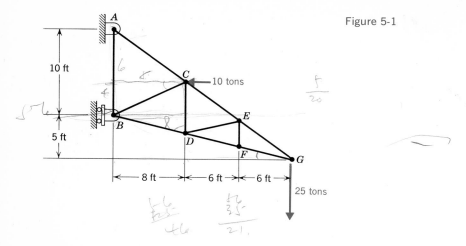

Figure 5-1

and corresponding equations of horizontal and vertical force equilibrium are as follows.

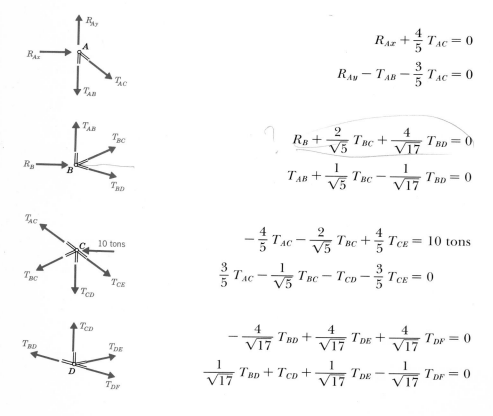

$$R_{Ax} + \frac{4}{5}\, T_{AC} = 0$$

$$R_{Ay} - T_{AB} - \frac{3}{5}\, T_{AC} = 0$$

$$R_B + \frac{2}{\sqrt{5}}\, T_{BC} + \frac{4}{\sqrt{17}}\, T_{BD} = 0$$

$$T_{AB} + \frac{1}{\sqrt{5}}\, T_{BC} - \frac{1}{\sqrt{17}}\, T_{BD} = 0$$

$$-\frac{4}{5}\, T_{AC} - \frac{2}{\sqrt{5}}\, T_{BC} + \frac{4}{5}\, T_{CE} = 10 \text{ tons}$$

$$\frac{3}{5}\, T_{AC} - \frac{1}{\sqrt{5}}\, T_{BC} - T_{CD} - \frac{3}{5}\, T_{CE} = 0$$

$$-\frac{4}{\sqrt{17}}\, T_{BD} + \frac{4}{\sqrt{17}}\, T_{DE} + \frac{4}{\sqrt{17}}\, T_{DF} = 0$$

$$\frac{1}{\sqrt{17}}\, T_{BD} + T_{CD} + \frac{1}{\sqrt{17}}\, T_{DE} - \frac{1}{\sqrt{17}}\, T_{DF} = 0$$

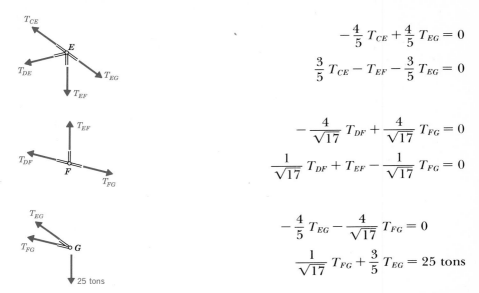

$$-\frac{4}{5}\,T_{CE} + \frac{4}{5}\,T_{EG} = 0$$

$$\frac{3}{5}\,T_{CE} - T_{EF} - \frac{3}{5}\,T_{EG} = 0$$

$$-\frac{4}{\sqrt{17}}\,T_{DF} + \frac{4}{\sqrt{17}}\,T_{FG} = 0$$

$$\frac{1}{\sqrt{17}}\,T_{DF} + T_{EF} - \frac{1}{\sqrt{17}}\,T_{FG} = 0$$

$$-\frac{4}{5}\,T_{EG} - \frac{4}{\sqrt{17}}\,T_{FG} = 0$$

$$\frac{1}{\sqrt{17}}\,T_{FG} + \frac{3}{5}\,T_{EG} = 25 \text{ tons}$$

The three support reaction components and 11 forces in the members can be determined from the above 14 equations of equilibrium.

The procedure illustrated above always leads to a set of independent algebraic equations for the unknown forces. If the truss is proper and statically determinate, the number of resulting equations will equal the number of unknown force components. In some cases, the set of equations can be solved readily, but in other cases a computer may be required to complete the solution. An efficient method of solving the above 14 equations is not obvious at first glance, so it may be best here to consider free-bodies of different portions of the structure, as is explained below.

Equations from Sections. The forces external to *any* portion of a system in equilibrium have zero resultants of force and moment. Often we can use this fact to obtain an equivalent, but much simpler set of equations.

To illustrate, reconsider the truss of Figure 5-1. A free-body diagram of the entire unit *ABG* is shown in Figure 5-2a. Summation of moments about point *A* is written as

$$R_B(10 \text{ ft}) - (10 \text{ tons})(6 \text{ ft}) - (25 \text{ tons})(20 \text{ ft}) = 0$$

from which

$$R_B = 56 \text{ tons} = 498 \text{ kN}$$

Then, summation of horizontal force components gives

$$R_{Ax} = -46 \text{ tons} = -409 \text{ kN}$$

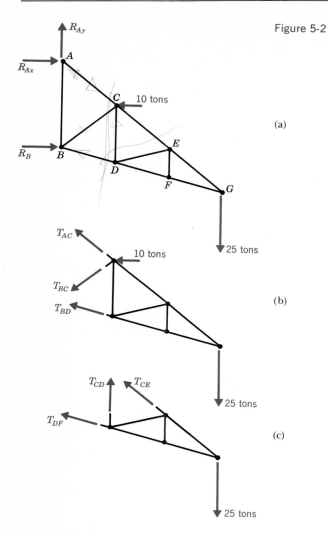

Figure 5-2

(a)

(b)

(c)

and summation of vertical force components gives

$$R_{Ay} = 25 \text{ tons} = 222 \text{ kN}$$

Next, for the free-body shown in Figure 5-2b, we can sum moments about point B:

$$\frac{4}{5} T_{AC} (10 \text{ ft}) + (10 \text{ tons})(4 \text{ ft}) - (25 \text{ tons})(20 \text{ ft}) = 0$$

to obtain

$$T_{AC} = 57.5 \text{ tons} = 512 \text{ kN}$$

Then we can sum moments about point G:

$$\frac{1}{\sqrt{5}} T_{BC} (12 \text{ ft}) + \left(\frac{2}{\sqrt{5}} T_{BC} + 10 \text{ tons}\right) (9 \text{ ft}) = 0$$

to obtain

$$T_{BC} = -3 \sqrt{5} \text{ tons} = 59.7 \text{ kN}$$

and sum moments about point C:

$$-\frac{4}{\sqrt{17}} T_{BD} (6 \text{ ft}) - (25 \text{ tons})(12 \text{ ft}) = 0$$

to obtain

$$T_{BD} = -\frac{25 \sqrt{17}}{2} \text{ tons} = 459 \text{ kN}$$

Now, considering the free-body in Figure 5-2c, we can obtain T_{CD} by summing moments about point G:

$$T_{CD} = 0$$

By continuing in this fashion, the remaining forces can be obtained without dealing with sets of simultaneous equations.

$$T_{AB} = -9.5 \text{ tons} = -84.5 \text{ kN}$$
$$T_{EG} = T_{CE} = 62.5 \text{ tons} = 556 \text{ kN}$$
$$T_{DE} = T_{EF} = 0$$
$$T_{FG} = T_{DF} = -\frac{25 \sqrt{17}}{2} \text{ tons} = -459 \text{ kN}$$

The strategy in this method is to isolate a portion of the structure, with the separation dividing the member in which a force is sought; then, with the free-body completed, to find a direction for force reckoning or an axis for moment reckoning that will yield an equation with as few unknown forces as possible.

For further illustration, consider the truss shown in Figure 5-3a. The vertical triangles ABC and DEF are both equilateral, and the rectangle $ABED$ is in a horizontal plane. The force in each of the nine members is to be determined.

Consider first the free-body of the portion of the truss shown in Figure 5-3b. To analyze the equilibrium of this section, we select various axes, about which moments involve only one unknown force. The moment about each axis is evaluated most readily by resolving each force as indicated in the illustration accompanying Equation 3-25c. Note that the moment of a force about an axis is zero if the line of action intersects the axis.

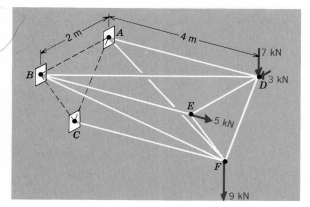

Figure 5-3

(a)

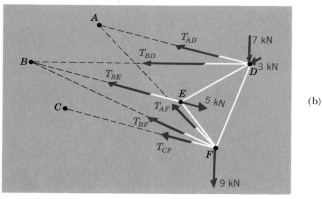

(b)

$$M_{DF} = (T_{BE} - 5 \text{ kN})(\sqrt{3} \text{ m}) = 0$$
$$T_{BE} = 5 \text{ kN}$$

$$M_{CF} = \left(\frac{2}{\sqrt{20}} T_{BD} + 3 \text{ kN}\right)(\sqrt{3} \text{ m}) - (7 \text{ kN})(1 \text{ m}) = 0$$

$$T_{BD} = \sqrt{5} \left(\frac{7}{\sqrt{3}} - 3\right) \text{ kN} = 2.33 \text{ kN}$$

$$M_{FE} = \left(T_{AD} + \frac{4}{\sqrt{20}} T_{BD}\right)(\sqrt{3} \text{ m}) = 0$$

$$T_{AD} = -2\left(\frac{7}{\sqrt{3}} - 3\right) \text{ kN} = -2.08 \text{ kN}$$

$$M_{BE} = \left(\frac{2}{\sqrt{20}} T_{AF}\right)(\sqrt{3} \text{ m}) - (9 \text{ kN})(1 \text{ m}) - (7 \text{ kN})(2 \text{ m}) = 0$$

$$T_{AF} = 23 \sqrt{\frac{5}{3}} \text{ kN} = 29.69 \text{ kN}$$

$$M_{DA} = \left(\frac{2}{\sqrt{20}} T_{BF}\right) (\sqrt{3} \text{ m}) - (9 \text{ kN})(1 \text{ m}) = 0$$

$$T_{BF} = 9 \sqrt{\frac{5}{3}} \text{ kN} = 11.62 \text{ kN}$$

$$M_{BA} = (T_{CF}) (\sqrt{3} \text{ m}) + (7 \text{ kN}) (4 \text{ m}) + (9 \text{ kN}) (4 \text{ m}) = 0$$

$$T_{CF} = -\frac{64}{\sqrt{3}} \text{ kN} = -36.95 \text{ kN}$$

To determine the forces in the remaining three members, it is a straight-forward matter to isolate joints D and E and sum forces.

$$T_{DF} = -\frac{14}{\sqrt{3}} \text{ kN} = -8.08 \text{ kN}$$

$$T_{DE} = T_{EF} = 0$$

Problems

5-1 For the truss in Figure 5-1, verify that the values of forces obtained from equations of sections satisfy the equations of joints.

5-2 For the truss in Figure 5-3, write the equations of equilibrium of each isolated joint. Verify that the forces obtained in the text satisfy these.

5-3 A snow load transfers the forces shown to each of the upper joints of a Pratt roof truss. Neglecting horizontal components of reactions at the supports, evaluate the forces in members BC, BF, CF, FG, and CG.

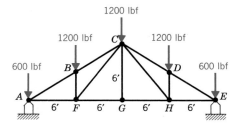

5-4 Evaluate the force in member DB.

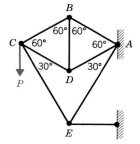

5-5 Evaluate the forces in members *CF* and *CG*.

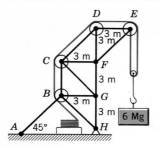

5-6 Evaluate the reactions and the forces in each member.

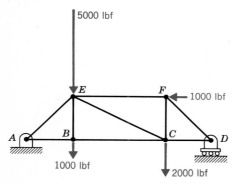

5-7 Evaluate the force in member *CD*. Members are all of equal length.

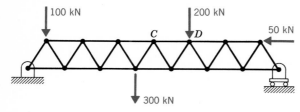

For Problems 5-8 through 5-12, evaluate the forces in those members assigned by your instructor.

5-8

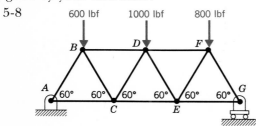

5-9

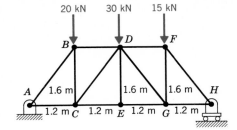

5-10

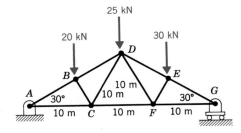

5-11

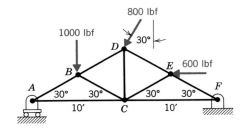

5-12

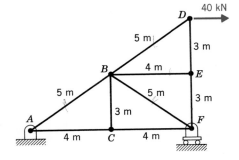

5-13   Determine the forces in members DC and CF, for

(a) $P_1 = 1$, $P_2 = P_3 = 0$

(b) $P_2 = 1$, $P_3 = P_1 = 0$

(c) $P_3 = 1$, $P_1 = P_2 = 0$
(d) Arbitrary P_1, P_2, P_3.

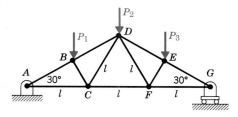

5-14 The members AD, CD, and BD, and the lines AB, BC, and CA are all of equal length. Evaluate the forces in the three members of the truss.

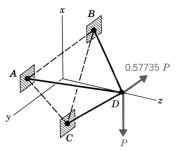

5-15 The members BF, FE, EC, ED, and FD, and the lines AB, BC, and CA are all of equal length, as are the members AF, AE, BD, and DC. BF and CE are perpendicular to ABC. Evaluate the forces in the members of the truss.

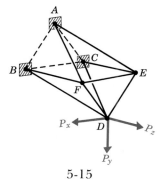

5-15

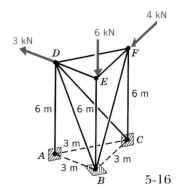

5-16

5-16 Members AD, BE, and CF are perpendicular to the plane ABC. Evaluate the forces in members BF, AD, and ED.

5–2

COUPLE-SUPPORTING MEMBERS.   The loads that a stiff bar can carry are not limited to axial forces. If lateral forces are applied, equilibrium requires that couples exist at various sections along the rod, as is demonstrated in Figure 5-4. This figure shows free-body diagrams of the beam and of two separated portions of the beam. Equilibrium of the portion on the left indicates at once that the forces across the plane of separation must form a couple and a lateral force.

The detailed distribution of interaction forces between the two parts of the beam is fairly complicated and cannot be deduced from equilibrium alone. (This is another subject, and is treated in the mechanics of deformable solids.) However, if the support system is statically determi-

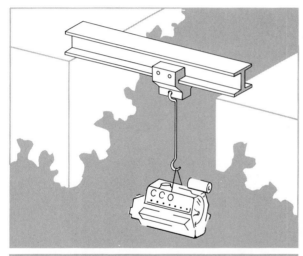

Figure 5-4

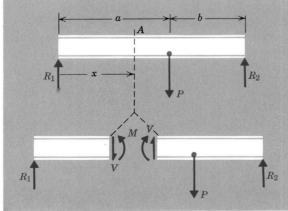

nate, it is possible to evaluate the moment of the equivalent couple. More-over, since structural failure is normally related directly to such moments, their evaluation is usually the most important aspect of the force analysis of a laterally loaded rigid bar.

Twisting and Bending Moments.   In general, the moment of the couple at a section can have any direction, depending on how the external loads are applied. For example, the bracket shown in Figure 5-5 has moment

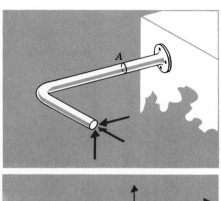

Figure 5-5

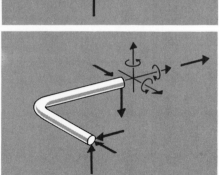

components in all three directions induced at section A. The moment vector is normally resolved into a component parallel to the axis of the bar and one or two components perpendicular to the axis. (This will facili-tate the further analysis of the strength and deformation of the bar.) The component of moment parallel to the axis of the bar is called the *twisting moment,* and the components perpendicular to the axis are called the *bending moments,* after the types of deformations they produce.

The resultant force acting at a section of a bar is similarly resolved. The component along the axis of the rod is called the *axial force,* and the components perpendicular to the axis are called the *shearing forces.*

Evaluation of these force and moment components is accomplished by using the same basic ideas already examined: a free-body of a portion

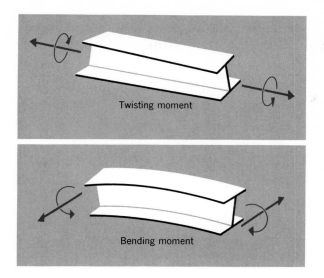

Twisting moment

Bending moment

of the member on either side of the section of interest is isolated and properly labeled, and equations of equilibrium are written and solved.

For example, equilibrium of the portion of the beam to the left of the section A in Figure 5-4 yields

$$V = R_1 \quad \text{and} \quad M = R_1 x$$

and equilibrium of the entire beam gives values of the support reactions in terms of the applied load P, as

$$R_1 = \frac{bP}{a+b} \qquad R_2 = \frac{aP}{a+b}$$

Thus, in terms of the applied load, the shear and bending moment are

$$V = \frac{bP}{a+b} \qquad x < a$$

$$M = \frac{bPx}{a+b} \qquad x < a$$

The qualification $x < a$ is necessary because the analysis was done for sections to the left of the applied load. A similar analysis for sections on the other side of the load results in

$$V = \frac{-aP}{a+b} \qquad x > a$$

$$M = \frac{a(a+b-x)P}{a+b} \qquad x > a$$

As another example, let us evaluate the force and moment components at the support O for the automobile torsion bar in Figure 5-6a.

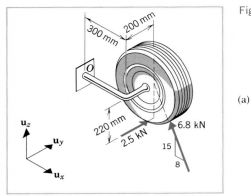

Figure 5-6

(a)

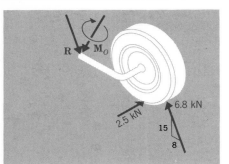

(b)

Force equilibrium of the free-body in Figure 5-6b requires that

$$\mathbf{R} + 6.8 \text{ kN} \left(-\frac{8}{17}\,\mathbf{u}_x + \frac{15}{17}\,\mathbf{u}_z \right) + 2.5 \text{ kN } \mathbf{u}_y = 0$$

Or

$$\mathbf{R} = 3.2 \text{ kN } \mathbf{u}_x - 2.5 \text{ kN } \mathbf{u}_y - 6.0 \text{ kN } \mathbf{u}_z$$

Therefore, there is a compressive axial force of 3.2 kN and a resultant shearing force of $\sqrt{(2.5)^2 + (6.0)^2}$ kN = 6.5 kN at the section O. Next, moment equilibrium requires that

$$\mathbf{M}_O + [(300 \text{ mm})\mathbf{u}_x + (200 \text{ mm})\mathbf{u}_y] \times [(6.0 \text{ kN})\mathbf{u}_z - (3.2 \text{ kN})\mathbf{u}_x]$$
$$+ [(300 \text{ mm})\mathbf{u}_x + (200 \text{ mm})\mathbf{u}_y - (220 \text{ mm})\mathbf{u}_z] \times (2.5 \text{ kN})\mathbf{u}_y = 0$$

Or,

$$\mathbf{M}_O = (-1.75 \text{ N·m})\mathbf{u}_x + (1.80 \text{ N·m})\mathbf{u}_y - (1.39 \text{ N·m})\mathbf{u}_z$$

This indicates a twisting moment,

$$M_t = (1.75) \text{ N·m}$$

and a bending moment resultant,

$$M_b = \sqrt{(1.80)^2 + (1.39)^2} \ \text{N·m}$$
$$= 2.27 \ \text{N·m}$$

Problems

5-17 Write expressions for shear and bending moment in sections to the right of the applied force in Figure 5-4.

5-18 Evaluate the shear force, the bending moment, and the twisting moment for each section of pipe shown in Problem 4-10. Draw diagrams that show how these vary along the pipe sections.

5-19 Evaluate the axial and shearing forces and the bending moment in the horizontal bar in Problem 4-15; show how these vary along the bar.

5-20 Evaluate the bending moment in the frame of Problem 4-18, and sketch its variation along the frame.

5-21 Evaluate the bending and twisting moments in each section of the shaft shown in Problem 4-29.

5-22 Evaluate the bending moment in the load leveler bar *AB* in Problem 4-30.

5-23 Evaluate the bending moment along the mast of Problem 4-32.

5-24 Plot the bending moment along the entire length of the frame.

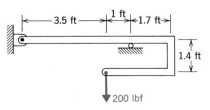

5-25 Plot the bending moment along the horizontal portion of the beam.

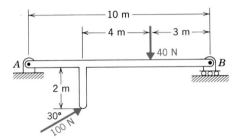

5-26 Plot the bending moment along the entire davit of Problem 4-17.

5–3

SYSTEMS WITH FRICTION.   Friction forces are those that act tangential to the surface on which two objects make contact. People have gone to considerable effort to minimize them in machinery of all kinds, but depend heavily on them at contact surfaces like those between shoe soles and floors, tires and roadways, and drive belts and pulleys.

Coulomb's Friction Law.   The nature of friction forces is a complicated subject that has been studied extensively. However, for most dry surfaces, the approximation of "Coulomb friction" has been found to lead to predictions of acceptable accuracy. The so-called Coulomb's law* of friction

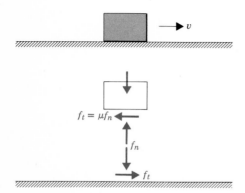

states that whenever sliding takes place the tangential component of force between two sliding surfaces is proportional to the normal component and acts in a direction to oppose the motion. This can be expressed analytically as

$$\mathbf{f}_t = -\mu f_n \mathbf{u}_v \qquad (5\text{-}1a)$$

in which $\mathbf{u}_v$ is a unit vector in the direction of the relative velocity of the object on which $\mathbf{f}_t$ acts, and the coefficient of friction μ depends on the surfaces in contact but not on the magnitude of the normal force nor on the velocity.

When the body is at rest, the friction force can have any direction required for equilibrium, but its magnitude is limited by

$$f_t \leq \mu f_n \qquad (5\text{-}1b)$$

When applied forces induce a friction force that reaches this limit, motion is incipient, meaning that any further increase will cause acceleration.

---

* For Charles August de Coulomb, whose early work on statics and mechanics was concerned in part with friction.

Example

What force is required to move the 80-lb block up the 15° incline shown in Figure 5-7? Also, in the absence of P, will the block remain stationary on the incline? The coefficient of friction is 0.30.

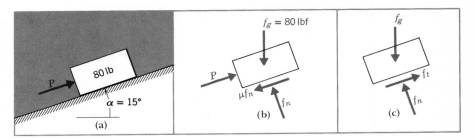

Figure 5-7

Referring to the free-body diagram in Figure 5-7b, we can write equations of equilibrium in the direction normal to the incline.

$$f_n - f_g \cos \alpha = 0$$

and in the direction along the incline:

$$P - f_g \sin \alpha - \mu f_n = 0$$

Elimination of f_n from these two equations results in

$$P = f_g (\sin \alpha + \mu \cos \alpha)$$
$$\doteq (80 \text{ lbf})(\sin 15° + 0.30 \cos 15°)$$
$$= 43.9 \text{ lbf}$$

The free-body diagram in Figure 5-7c depicts the situation in the absence of the force P. Observe the reversal of the direction of the friction force. Now, if the block is to remain at rest,

$$f_n = f_g \cos \alpha$$
$$f_t = f_g \sin \alpha$$

But the friction force is limited by

$$f_t \leq \mu f_n$$

Substitution of the equilibrium equations into this inequality gives

$$f_g \sin \alpha \leq \mu f_g \cos \alpha$$

Or,

$$\tan \alpha \leq \mu$$

But $\tan \alpha = 0.27$, which is less than $\mu = 0.30$; therefore, the block will remain at rest.

Some surfaces transmit a slightly smaller tangential force during slipping than is needed to start slipping from rest. When this difference is important, it is accounted for by using a "dynamic coefficient of friction" μ_d and a "static coefficient of friction" μ_s that is greater than μ_d. This modification of Coulomb's law is expressed analytically as follows:

$$\mathbf{f}_t = -\mu_d f_n \mathbf{u}_v \qquad \mathbf{v} \neq \mathbf{0} \tag{5-2a}$$
$$f_t \leq \mu_s f_n \qquad \mathbf{v} = \mathbf{0} \tag{5-2b}$$

Example

In the absence of P, the angle of the incline in Figure 5-7 is slowly increased until the block begins to slide downward. The static coefficient of friction is $\mu_s = 0.47$ and the dynamic coefficient of friction is $\mu_d = 0.44$. What will be the acceleration of the block after it breaks loose?

Prior to breakaway of the block,

$$f_t \leq \mu_s f_n$$

Or, introducing the equilibrium relationships (see Figure 5-7c),

$$f_g \sin \alpha \leq \mu_s f_g \cos \alpha$$

The equality in this relationship occurs when the critical angle α_c is reached:

$$\tan \alpha_c = \mu_s = 0.47 \tag{a}$$

After the block breaks loose,

$$f_t = \mu_d f_n \tag{b}$$

Now, the normal component of acceleration is zero;

$$f_n - f_g \cos \alpha_c = 0 \tag{c}$$

but in the tangential direction, we must replace the usual static equilibrium equation with Newton's second law,

$$f_g \sin \alpha_c - f_t = ma \tag{d}$$

in which a is the downward tangential acceleration. Substituting (b) and (c) into (d),

$$f_g \sin \alpha_c - \mu_d f_g \cos \alpha_c = ma$$

But since $f_g = mg$,

$$a = g(\sin \alpha_c - \mu_d \cos \alpha_c)$$
$$= g \sin \alpha_c \left(1 - \frac{\mu_d}{\tan \alpha_c} \right)$$

Substituting from (a),

$$a = g \sin \alpha_c \left(1 - \frac{\mu_d}{\mu_s}\right)$$

$$= (9.807 \text{ m/s}^2)(0.4254)\left(1 - \frac{0.44}{0.47}\right)$$

$$= 0.266 \text{ m/s}^2$$

Belt Friction. Many devices incorporate a flexible line passed around a solid circular object, such as a pulley, with torque transmitted by means of the tangential (friction) forces along the contact surface. In such a mechanism, the transmission of torque results in a difference in tension between the two ends of the line, the tension varying along the portion in contact in a manner that is revealed in the following analysis.

Figure 5-8a shows the essential features of the device while Figure 5-8b shows a free-body diagram of an element of the line. We assume that Coulomb's law is valid for the surfaces and that slipping is incipient.

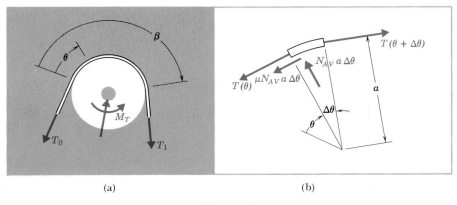

(a)   (b)

Figure 5-8

Equilibrium of the element in the radial direction is given by

$$N_{av}a \, \Delta\theta - T(\theta + \Delta\theta) \sin \Delta\theta = 0$$

in which N_{av} represents an average radial force per unit length of the line. Division by $\Delta\theta$ and evaluation of the limit as $\Delta\theta \to 0$ lead to

$$aN(\theta) = T(\theta)$$

Equilibrium in the tangential direction is given by

$$T(\theta + \Delta\theta) \cos \Delta\theta - T(\theta) - \mu N_{av}a \, \Delta\theta = 0$$

Division by $\Delta\theta$ and evaluation of the limit as $\Delta\theta \rightarrow 0$ lead to

$$\frac{dT}{d\theta} - \mu a N = 0$$

Elimination of N from the two equilibrium relationships gives

$$\frac{dT}{d\theta} - \mu T = 0$$

This may be rearranged in the form

$$\frac{dT}{T} = \mu d\theta$$

and integrated to give

$$\log T = \mu\theta + C$$

The constant of integration can be related to the tension T_0 by setting $\theta = 0$:

$$\log T_0 = C$$

Then, the tension-angle relationship becomes

$$\log \frac{T}{T_0} = \mu\theta$$

Or,

$$T = T_0 e^{\mu\theta}$$

This is valid for all values θ within the region of contact. With the total subtended angle of contact denoted by β, the tensions at the two points of tangency are related by

$$\frac{T_1}{T_0} = e^{\mu\beta} \tag{5-3}$$

From Figure 5-8a we see that the torque transmitted is given by

$$M_T = T_1 a - T_0 a$$
$$= T_0 a (e^{\mu\beta} - 1) \tag{5-4}$$

Example

How far around a tree must a wrangler wrap a rope to hold an excited horse by pulling with a force of 20 lb? The horse can pull with a force of 650 lb and the coefficient of friction between the rope and the tree is 0.3.

   If slipping is incipient, Equation 5-3 requires that

$$\frac{650 \text{ lbf}}{20 \text{ lbf}} = e^{0.30\beta}$$

Solving for the angle, we obtain

$$\beta = \frac{1}{0.30} \log 32.5$$
$$= 11.6 \text{ radians}$$
$$= 1.8 \text{ revolutions}$$

Problems

5-27   What is the value of the friction force when $\alpha = 15°$? What is the critical value of α at which the block will start sliding down the plane?

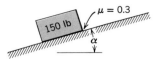

5-28   Let $\alpha = 20°$ in the system of the preceding problem. What will be the magnitude of the acceleration of the block as it slides down the plane? If it is sliding upward, having had an initial motion in this direction imparted to it, what will be the magnitude of its acceleration?

5-29   Show from the geometry of a force polygon that the relationship

$$\tan \alpha \leqslant \mu$$

limits the incline on which a friction-held body can remain at rest.

5-30   Will the crate slide down or tip over in the absence of P? The contents are such that the forces of gravity may be represented by a single force through the geometric center. What is the highest point at which P can be applied without tipping the crate?

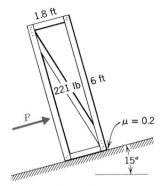

5-31   Rework Problem 5-30, with P horizontal rather than parallel to the incline.

5-32 Determine the minimum value of P necessary to start motion. The coefficient of friction between all surfaces is 0.27.

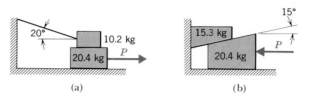

(a)   (b)

5-33 The ladder of mass m rests against the wall as shown. The coefficient of friction at both surfaces is μ. Determine how far up the ladder the man of mass M can walk before the ladder slips.

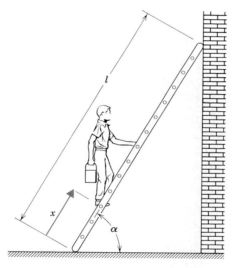

5-34 What coefficient of friction is necessary to prevent slipping between the yo-yo and the floor? Which way will the yo-yo roll, assuming there is sufficient friction to prevent slip?

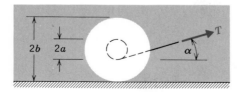

5-35 Determine the minimum coefficient of friction between the block and the bar necessary to prevent collapse.

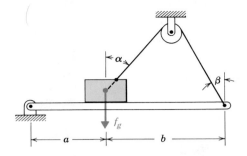

5-36 Determine the minimum coefficient of friction between the block and the bar necessary to prevent collapse.

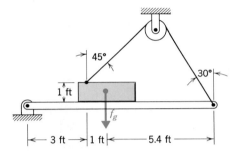

5-37 Determine the minimum coefficient of friction between the wheel and the step that will permit the wheel to roll over the step.

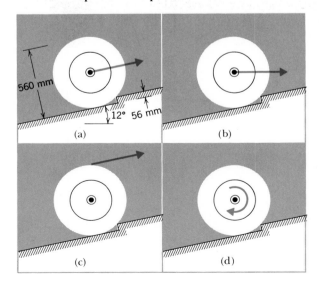

5-38   Until a clamp is tightened, the drill press table is free to slide along
the column. Estimate the coefficient of friction required so that the
collar will be self-locking against the column under the action of
thrust from the drill.

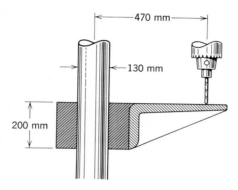

5-39   In terms of the axial force P, the coefficient of friction μ, and the
inside and outside radii a and b, what is the moment M necessary to
cause slipping. Assume that the normal pressure is uniform over
the surface of contact.

5-40   Estimate the coefficient of friction necessary in order that the oil
filter wrench does not slip.

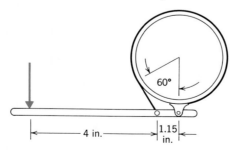

5-41   What braking force f is required to exert a friction moment of
115 N·m opposing clockwise rotation of the drum? The coeffi-

cient of friction is 0.2. What would be the required force if the rotation is counterclockwise?

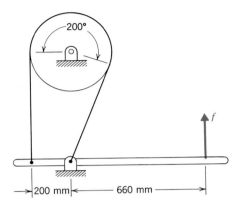

5-42   The lever arm in the differential band brake is arranged so that the tight side of the band aids the actuating force while the loose side opposes it. Show that the mechanism can be made so as to permit rotation in one direction only, while it is self-locking in the other direction. This type of device is commonly used on hoists to prevent a load from dropping.

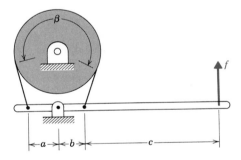

5-43   What is the ratio of the force f_2 to the force f_1 in the force amplifier?

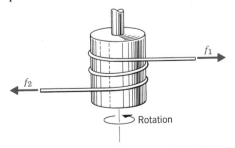

DISTRIBUTED FORCES

Each of the forces that we have considered thus far as being concentrated along a discrete line of action, is actually a resultant of a set of forces distributed throughout some region. With contact forces, the distribution is over some surface area; with gravitational forces, the distribution is throughout some volume. In effect, this amounts to a composition of the components within the distributed set of forces.

If the region is sufficiently small, or the distribution has a certain regularity, the composition can be made correctly by inspection. In this chapter, we are concerned with the computation of resultants of distributed forces in cases that do not lend themselves readily to composition by inspection.

The procedures are based on the same ideas that are explained in Section 3-5 and in Chapter 4; in handling a set of distributed forces the only new feature is the computational detail of summation, which takes the form of integration.

Example

The uniform, slender, semicircular arch shown in Figure 6-1 is acted on by gravity and the reactions from the roller supports. Evaluate the bending moment across the section at the top of the arch.

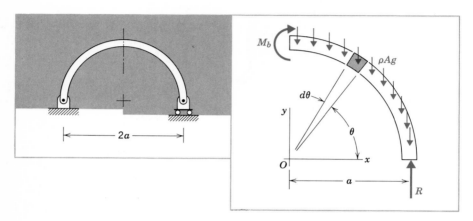

Figure 6-1

The free-body diagram shows the desired bending moment M_b, the reaction from the other half of the arch. Because of horizontal equilibrium no axial force exists at this section. The shearing force is also zero at this section, because Newton's third law would demand an oppositely directed shearing force on the other half of the arch, and this pair of forces would be inconsistent with the symmetry of the system.

Because the arch is slender, the forces of gravity may be treated as distributed along a circular *line*. Let the cross-sectional area be denoted by A and the density (mass per unit volume) by ρ. Then the volume of the shaded element of the arch will be equal to $Aa\,d\theta$, and the magnitude of the force of gravity acting on it will be given by

$$df_g = \rho(Aa\,d\theta)g$$

The resultant of the gravitational forces then has the magnitude

$$f_g = \int_0^{\pi/2} \rho Aga\,d\theta$$
$$= \tfrac{1}{2}\pi a\rho\,Ag$$

and vertical force equilibrium gives the support reaction as

$$R = \tfrac{1}{2}\pi a\rho Ag$$

Now, let us compute the resultant moment of the forces of gravity about the point O. The direction is into the plane of the sketch (clockwise) and the magnitude is given by

$$M_{Og} = \int a \cos \theta \, df_g$$

$$= \int_0^{\pi/2} a \cos \theta \, \rho A g a \, d\theta$$

$$= a^2 \rho A g$$

Finally, moment equilibrium requires that

$$Ra - M_{Og} - M_b = 0$$

which, together with the above results, gives

$$M_b = \left(\frac{1}{2} \pi - 1 \right) a^2 \rho A g$$

In an arch with a large radius, this bending moment could easily cause failure of the structure.

6–1

SINGLE FORCE EQUIVALENTS.   The setting up and evaluating of integrals such as in the example above can be circumvented if knowledge of an equivalent discrete force is available. This information, for a variety of special cases, is available in tabulated formulas such as appear in Appendix B of this volume. These formulas are generated by such integrations.

The concepts developed in this section provide the understanding necessary for proper use of such tables and are useful in applications other than statics.

Center of Mass and Center of Gravity.   Consider the forces of gravity distributed throughout an arbitrary body, as shown in Figure 6-2. If the body is small with respect to the earth, the forces will be parallel, and g will have the same value throughout the body. The force acting on the element with mass dm will be

$$d\mathbf{f}_g = dm \, g \, \mathbf{u}_g$$

and the resultant force will be given by

$$\mathbf{f}_g = \int d\mathbf{f}_g$$

$$= \int (dm \, g \, \mathbf{u}_g)$$

$$= \left(\int dm \right) g \, \mathbf{u}_g$$

$$= mg \, \mathbf{u}_g$$

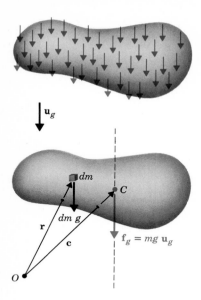

Figure 6-2

where m is the total mass of the body. The resultant moment about a point O is given by

$$\mathbf{M}_{Og} = \int \mathbf{r} \times d\mathbf{f}_g$$
$$= \int \mathbf{r} \times (g\,\mathbf{u}_g\,dm)$$
$$= (\int \mathbf{r}\,dm) \times g\,\mathbf{u}_g$$

where $\mathbf{r}$ is the position vector locating the mass element. Now, to compose the forces into a single resultant, the line of action of the equivalent force $\mathbf{f}_g$ must pass through a point located by the position vector $\mathbf{r}_f$, satisfying the moment equivalence

$$\mathbf{r}_f \times \mathbf{f}_g = \mathbf{M}_{Og}$$

Or,

$$\mathbf{r}_f \times (mg\,\mathbf{u}_g) = (\int \mathbf{r}\,dm) \times g\,\mathbf{u}_g$$
$$\mathbf{r}_f \times \mathbf{u}_g = \mathbf{c} \times \mathbf{u}_g$$

where

$$\boxed{\mathbf{c} = \frac{1}{m}\int \mathbf{r}\,dm} \tag{6-1}$$

The vector $\mathbf{c}$ locates an important point C, called the *center of mass* of the body. It is clear from the equation preceding (6-1) that by choosing $\mathbf{r}_f = \mathbf{c}$, moment equivalence will be satisfied for any $\mathbf{u}_g$, which implies that the line of action of the equivalent force passes through C regardless of how the body is oriented.

The center of mass C is of fundamental importance in the study of dynamics. In statics it significance lies in the fact that the resultant moment about C, of the force of uniform gravity, is zero. That is, the body could be statically balanced by supporting it at this point only. For this reason it is also called the center of gravity.*

Example

Find the location of the center of gravity of the portion of arch isolated as a free-body in Figure 6-1.

Using the center of the circle as a reference point, we can locate the shaded mass element with the position vector

$$\mathbf{r} = a \cos \theta \, \mathbf{u}_x + a \sin \theta \, \mathbf{u}_y$$

Then, from (6-1),

$$\mathbf{c} = \int \frac{1}{m} (a \cos \theta \, \mathbf{u}_x + a \sin \theta \, \mathbf{u}_y)(\rho \, Aa \, d\theta)$$

$$= \frac{1}{m} \rho Aa^2 \left[\left(\int_0^{\pi/2} \cos \theta \, d\theta \right) \mathbf{u}_x + \left(\int_0^{\pi/2} \sin \theta \, d\theta \right) \mathbf{u}_y \right]$$

$$= \frac{\rho A \, a^2 (\mathbf{u}_x + \mathbf{u}_y)}{\rho Aa \, \pi/2}$$

$$= \frac{2a}{\pi} \mathbf{u}_x + \frac{2a}{\pi} \mathbf{u}_y$$

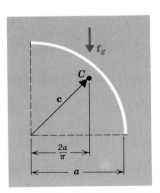

* If the gravitational field varies in magnitude or direction throughout the body, the forces of gravity can produce a moment about the center of mass. This small moment can be important to the rotational motion of bodies such as satellites and planets. In this case, it is not reasonable to refer to C as the center of gravity. However, the center of mass is a property of the body itself, independent of the environment.

Centroids. The mass dm of the element used in the preceding integrals can be expressed in terms of the density ρ and the corresponding element of volume dV, as $dm = \rho \, dV$. Then Equation 6-1 can be written as

$$\mathbf{c} = \frac{\int \mathbf{r}\rho \, dV}{\int \rho \, dV}$$

Now, if the density is uniform throughout the body, ρ can be brought outside the integrals, with the result

$$\boxed{\mathbf{c} = \frac{1}{V} \int \mathbf{r} \, dV} \tag{6-2}$$

The location of the point C here depends entirely on geometry, since all contribution having to do with material has been canceled. The point C, located according to (6-2), is called the *centroid of the volume V*.

Similarly, the *centroid of a surface area A* is defined as the point located by the position vector

$$\boxed{\mathbf{c} = \frac{1}{A} \int \mathbf{r} \, dA} \tag{6-3}$$

and the *centroid of a line segment* of length L is defined as the point located by the position vector

$$\boxed{\mathbf{c} = \frac{1}{L} \int \mathbf{r} \, dL} \tag{6-4}$$

The calculation in the preceding example was for a uniform mass per unit length of the arch; this density having canceled out, the center of mass of the segment of arch is also the centroid of a quarter segment of a circular line.

The integrals $\int \mathbf{r} \, dV$, $\int \mathbf{r} \, dA$, and $\int \mathbf{r} \, dL$ are called the first moments of volume, area, and line, respectively, about the reference point O.

In carrying out the calculations, it is usually convenient to work with one rectangular Cartesian component of the position vectors at a time, that is, to evaluate separately the component equivalents of (6-2). These are, in terms of $\mathbf{r} = x\mathbf{u}_x + y\mathbf{u}_y + z\mathbf{u}_z$,

$$c_x = \frac{1}{V} \int x \, dV$$

$$c_y = \frac{1}{V} \int y \, dV$$

$$c_z = \frac{1}{V} \int z \, dV$$

Example

Find the location of the centroid of the wedge-shaped volume shown in Figure 6-3.

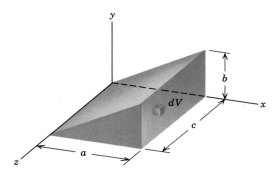

Figure 6-3

In terms of the rectangular Cartesian coordinates shown, the volume element can be expressed as

$$dV = dx\, dy\, dz$$

and the equation for the slanted surface is

$$x = \frac{ay}{b}$$

The integrals needed for determing c_x are evaluated next as follows:

$$V = \int_0^c \int_0^b \int_{ay/b}^a dx\, dy\, dz$$

$$= \int_0^c \int_0^b a\left(1 - \frac{y}{b}\right) dy\, dz$$

$$= \int_0^c a\left(b - \frac{b^2}{2b}\right) dz$$

$$= \frac{1}{2}\, abc$$

$$\int x\, dV = \int_0^c \int_0^b \int_{ay/b}^a x\, dx\, dy\, dz$$

$$= \int_0^c \int_0^b \frac{1}{2}\left[a^2 - \left(\frac{ay}{b}\right)^2\right] dy\, dz$$

$$= \int_0^c \frac{1}{2}\, a^2\left(b - \frac{b^3}{3b^2}\right) dz$$

$$= \frac{1}{3}\, a^2bc$$

Then,

$$c_x = \frac{1}{V} \int x \, dV$$

$$= \frac{\frac{1}{3}a^2 bc}{\frac{1}{2}abc}$$

$$= \frac{2a}{3}$$

Similar calculations lead to the other coordinates of the centroid,

$$c_y = \frac{b}{3}$$

$$c_z = \frac{c}{2}$$

A composite volume, area, or line may be made up from several parts, of which each centroid is known. In this case a summation having the same form as the integral definitions can be readily shown to be valid. In the case of an area A that is a composite formed from several areas A_1, A_2, . . . , having centroids located by c_1, c_2, . . . , the centroid of the composite would be located by

$$\mathbf{c} = \frac{\mathbf{c}_1 A_1 + \mathbf{c}_2 A_2 + \cdot \cdot \cdot}{A_1 + A_2 + \cdot \cdot \cdot}$$

Example

Locate the centroid of the plane area shown in Figure 6-4.

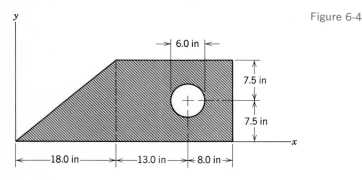

Figure 6-4

The areas and coordinates of centroids of individual parts are as follows:

	A, in^2	c_x, in	c_y, in
Triangle	135	12.0	5.0
Rectangle	315	28.5	7.5
Circle	−28.3	31.0	7.5

The coordinates locating the centroid are then

$$c_x = \frac{(12.0)(135) + (28.5)(315) + (31.0)(-28.3)}{135 + 315 - 28.3}$$

$$= 23.05 \text{ in}$$

$$c_y = \frac{(5.0)(135) + (7.5)(315) + (7.5)(-28.3)}{135 + 135 - 28.3}$$

$$= 6.70 \text{ in}$$

Problems

6-1 The mass of the earth is about 5.98×10^{24} kg, and that of the moon about 7.35×10^{22} kg. The center-to-center distance between the earth and the moon is 3.84×10^5 km (mean). The radius of the earth is 6380 km. With respect to the surface of the earth, where is the center of mass of the earth-moon system?

6-2 At what angle α will the plate hang, suspended at rest?

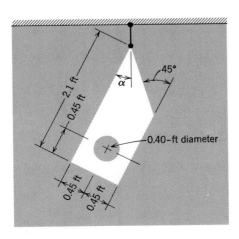

6-3   Locate the center of mass of the system consisting of the three homogeneous blocks.

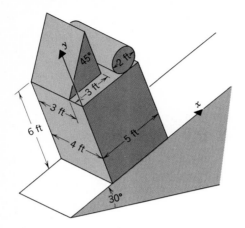

6-4   Determine the tension in each of the cables supporting the uniform slab.

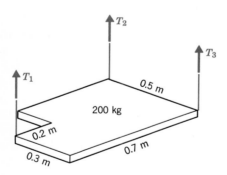

6-5   Verify the following formulas in Appendix B-2:
(a) c_x and c_y for the triangle
(b) c_x and c_y for the area bounded by the x axis and the curve $y = b x^n/a^n$.
(c) c_x and c_y for the segment of a circle
(d) c_x for the segment of a sphere
(e) c_x, c_y, and c_z for the cone

6-6   Locate the line of action of the force equivalent to the gravitational forces acting on the concrete dam in Figure 6-6.

6-7 The 100-mm diameter sphere has a 40-mm hole drilled radially to its center, the bottom of the hole being flat. Locate the center of mass.

6–2

SUBMERGED BODIES. Submerged structures are usually required to withstand significant forces from surrounding fluid. The simplest to predict are those that are induced by gravity when the fluid is at rest. The hydrostatic loading then acts normally to every surface and varies only with depth; from these facts we can develop some special techniques for computing force and moment resultants.

Static Fluid Pressure. When a fluid is at rest, it is capable of transmitting contact forces only in a direction perpendicular to the surface of contact. Here we refer not only to the interfaces between the fluid and some object but to any surface separating different parts of a fluid-filled region. This property distinguishes fluids from solids.

   The intensity of this contact force, or force per unit of area, is called *pressure*. In general, the force intensity varies from point to point within the region, so that a definition of pressure at a point involves a limit, as the area considered approaches zero, of the ratio of force to area. In the SI system the unit of pressure is the newton per square meter, called the pascal (Pa). Atmospheric pressure at sea level is about 10^5 Pa $= 100$ kPa.

   A consequence of the fact that the contact force acts normal to every surface is that the pressure at a point is the same in every direction. To show this, consider the free-body of the element of fluid in Figure 6-5. The area of the inclined surface is denoted as ΔA, and we suppose it to be sufficiently small that variations of pressure over each of the four surfaces may be neglected. The orientation of the inclined surface is specified by the unit vector normal to the surface:

$$\mathbf{u}_n = n_x\mathbf{u}_x + n_y\mathbf{u}_y + n_z\mathbf{u}_z$$

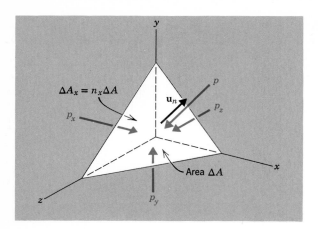

Figure 6-5

That is, n_x, n_y, and n_z are the direction cosines of the normal to the surface. The areas of the three surfaces perpendicular to the axes are then (See Problem 3-15.)

$$\Delta A_x = n_x\,\Delta A, \quad \Delta A_y = n_y\,\Delta A, \quad \Delta A_z = n_z\,\Delta A$$

The volume of the element is $\Delta V = \frac{1}{3}\,\Delta A\,h$, where h is the perpendicular distance from the origin to the inclined surface. The force of gravity then has the magnitude $\frac{1}{3}\,\Delta A\,h\rho g$. With the x component of this force denoted as $\frac{1}{3}\,\Delta A\,h\rho g_x$, the equilibrium of forces in the x direction gives

$$p_x(n_x\,\Delta A) - (p\,\Delta A)n_x + \frac{1}{3}\,\Delta A\,h\rho g_x = 0$$

Or,

$$(p_x - p)n_x + \frac{1}{3}\,h\rho g_x = 0$$

Now, as we let $h \to 0$, this reduces to

$$p_x = p$$

The same consideration in the other two directions gives

$$p_y = p_z = p$$

Thus the pressure at a point in a fluid at rest is the same in every direction.

We can make a useful interpretation of force relationships described above. The x component of the force acting on the inclined surface, $(p\,\Delta A)n_x$, is equal to the force $p(n_x\,\Delta A)$ acting on the surface perpendicular to the x axis. But this latter area is the projection of the inclined surface onto the plane perpendicular to the x direction. Therefore, the component in any direction, of the force acting on an arbitrary element is equal to the product of the pressure and the projection of the element area onto the plane perpendicular to this direction. This *projected area interpretation* often simplifies the integration to obtain a component of the resultant of pressure over nonplanar surfaces.

An *incompressible* fluid is an idealization in which the density of the fluid is independent of pressure. Liquid water will undergo an increase in density of only 0.0023 % under the influence of 10 000 atmospheres of pressure, so it is essentially incompressible for our present purposes.

In a body of incompressible fluid, gravity will induce pressure that increases linearly with depth. To verify this and determine the rate of increase, consider a free-body of a vertical cylindrical portion of fluid with cross sectional area ΔA and height $z_2 - z_1$. The horizontal forces are missing from the free-body in the sketch. However, we can deduce from the projected area interpretation that these forces are self-canceling provided that the pressure varies only with z and not in the horizontal directions. Vertical force equilibrium requires that

$$f_g = \rho \, \Delta A (z_2 - z_1) g = p_2 \, \Delta A - p_1 \Delta A$$

Or,

$$p_2 - p_1 = \rho g (z_2 - z_1) \tag{6-5}$$

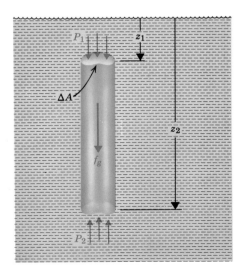

For a container of arbitrary shape, the pressure change can be traced anywhere by using connected elements as above; therefore, (6-5) is valid for any two points that are connected by a region filled with the particular incompressible fluid.

For fresh water near the earth's surface,

$$\rho g = 9.8 \text{ kPa/m} = 62.4 \text{ lbf/ft}^3$$

Force Resultants.

Example

Evaluate the resultant force transmitted by the water to the dam depicted in Figure 6-6.

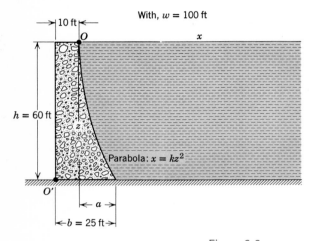
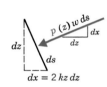

Figure 6-6

The force acting on the element of dam surface at depth z can be expressed as

$$d\mathbf{f} = p(z)w \, ds \left(\frac{dx}{ds} \mathbf{u}_z - \frac{dz}{ds} \mathbf{u}_x \right)$$

$$= p(z)w \, (dx \, \mathbf{u}_z - dz \, \mathbf{u}_x)$$

Notice that this last expression could be written directly by using the projected area interpretation. With atmospheric pressure denoted by p_a, the pressure at depth z is

$$p(z) = p_a + \rho gz$$

By using this and the differential relationship $dx = 2kz \, dz$ above, we can form the integral expression for the resultant force:

$$\mathbf{f} = w \int_0^h (p_a + \rho gz)(2kz \, \mathbf{u}_z - \mathbf{u}_x) \, dz$$

Use of the relationship $a = kh^2$ and evaluation of the integrals lead to

$$f_x = -wh\left(p_a + \frac{1}{2}\rho gh\right) \tag{a}$$

$$f_z = wa\left(p_a + \frac{2}{3}\rho gh\right) \tag{b}$$

Finally, substitution of the numerical values

$$p_a = 2100 \text{ lbf/ft}^2$$
$$\rho g = 62.4 \text{ lbf/ft}^3$$
$$w = 100 \text{ ft}$$
$$h = 60 \text{ ft}$$
$$a = 25 \text{ ft} - 10 \text{ ft} = 15 \text{ ft}$$

results in

$$f_x = -23.8 \times 10^6 \text{ lbf}$$
$$f_z = 6.9 \times 10^6 \text{ lbf}$$

Equations (a) and (b) can be obtained from a different point of view from that based on the force element shown in Figure 6-6. Consider the free-body of the portion of water shown in Figure 6-7. By Newton's third

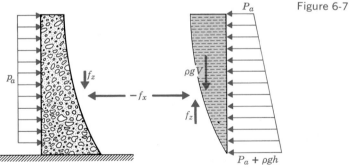

Figure 6-7

law, the force acting on the curved surface is of equal magnitude but has opposite direction from the force we wish to evaluate. Considering horizontal force equilibrium of this free-body, we can write

$$-f_x = \int_0^h p(z)w \, dz$$

Now note that from the viewpoint of the geometry of the pressure distribution diagram, this is also equal to the volume under the pressure "envelope." In terms of well-known formulas for the areas of rectangles and triangles and the volume within a cylinder, we can write

$$-f_x = \left(p_a h + \frac{1}{2}\rho g h^2\right) w \qquad [a]$$

Next, considering vertical force equilibrium, we have

$$f_z = p_a a w + \rho g V$$

where V is the volume of water in the free-body. This volume can be evaluated as

$$V = \int_0^h (a - x)w\,dz$$
$$= \int_0^h (a - kz^2)\,dz$$
$$= \frac{2}{3}wah$$

leading to the result (b).

Observe that the above calculations are evaluations of the force resultant that the *water* applies to the upstream surface of the dam. Other forces acting on the dam are from atmospheric pressure over the top and the downstream side, and the components of reaction across the bottom and ends.

Consider the effect of the air pressure. The density of air at sea level and 0° C is about 1.3 kg/m³, so that the pressure will be approximately

$$\rho_a g h = (1.3 \text{ kg/m}^3)(9.8 \text{ N/kg})(18.3 \text{ m})$$
$$= 0.23 \text{ kPa}$$
$$= 0.0023 \, p_a$$

greater at the bottom of the dam than at the top. With this difference neglected, the pressure over the downstream surface yields a resultant force of magnitude $p_a wh$. When all horizontal components acting on the dam are summed, this force cancels that expressed by the first term in equation (a). Therefore, the horizontal component of reaction at the base of the dam can be calculated without regard for the air pressure. The remaining effect of the atmospheric pressure is to induce an increased vertical component of reaction at the base.

Archimedes' principle. Consider a submerged object, with its underneath surface in contact only with the fluid. Because of the pressure gradient induced by gravity, the upward acting fluid pressure on the underneath surface will be greater than the downward acting pressure on the upper surface. The net effect is an upward force called the *buoyant force* acting on the object. Now if the object were to be replaced by fluid identical to that surrounding the object, the buoyant force acting on this replacement fluid would be the same as that on the original object. But since the replacement fluid would be in equilibirum, *the magnitude of this buoyant force must be identical with the magnitude of the force of gravity on the displaced fluid.* This is known as Archimedes' Principle.

Example

A sphere having a density ρ_s and radius a is released at the surface of a body of still water that has a density ρ. At what depth will the sphere reach equilibrium?

The volume of water displaced is equal to that of the spherical segment bounded by the plane of the water surface and the portion of the sphere below the surface. In terms of depth h, this volume is

$$V_{\text{displ}} = \pi h^2 \left(a - \frac{h}{3} \right) \qquad 0 \leqslant h \leqslant 2a$$

Therefore, the magnitude of the buoyant force is

$$f_b = \rho g \pi h^2 \left(a - \frac{h}{3} \right) \qquad 0 \leqslant h \leqslant 2a$$

and if the sphere is to be in equilibrium, this must be equal to the magnitude of the force of gravity acting on the sphere:

$$f_b = f_g$$

$$\rho g \pi h^2 \left(a - \frac{h}{3} \right) = \rho_s g \left(\frac{4\pi a^3}{3} \right) \qquad 0 \leqslant h \leqslant 2a$$

Or,

$$\left(\frac{h}{a} \right)^3 - 3 \left(\frac{h}{a} \right)^2 + \frac{4\rho_s}{\rho} = 0 \qquad 0 \leqslant \frac{h}{a} \leqslant 2$$

The root of this cubic equation that lies within the region $0 \leqslant h/a \leqslant 2$ has the value

$$\frac{h}{a} = 1 + 2 \cos \left(\frac{\phi + 4\pi}{3} \right)$$

where ϕ is given by

$$\cos \phi = 1 - \frac{2\rho_s}{\rho}$$

What happens when $\rho_s > \rho$?

Moment Resultants.

Example

Reconsider the dam depicted in Figure 6-6. Referring to the element shown there, we can compute the resultant moment of the hydrostatic forces about the y axis (which is directed outward from the plane of the sketch) as follows:

$$\mathbf{M}_{Oy}\text{ (water)} = -\mathbf{u}_y \int p(z)w(x\ dx + z\ dz)$$

$$= -\mathbf{u}_y w \int_0^h (p_a + \rho gz)[kz^2(2kz) + z]dz$$

$$= -\mathbf{u}_y w \int_0^h (p_a + \rho gz)\left(\frac{2a^2z^3}{h^4} + z\right)dz$$

$$= -\mathbf{u}_y w \left[\frac{1}{2}p_a(a^2 + h^2) + \rho gh\left(\frac{1}{3}h^2 + \frac{2}{5}a^2\right)\right]$$

Similarly, the resultant moment of the atmospheric pressure across the top surface and downstream side is

$$\mathbf{M}_{Oy}\text{(air)} = \mathbf{u}_y p_a w\ \frac{(b-a)^2 + h^2}{2}$$

The moment about the y axis of all forces of contact with air and water above the base is then

$$\mathbf{M}_{Oy} = \mathbf{M}_{Oy}\text{(water)} + \mathbf{M}_{Oy}\text{(air)}$$

$$= -\mathbf{u}_y w \left[p_a b\left(a - \frac{1}{2}b\right) + \rho gh\left(\frac{1}{3}h^2 + \frac{2}{5}a^2\right)\right]$$

The moment of these same forces about a parallel axis through the point O' can be evaluated by similar calculation or by using the above result and (3-28). The result is

$$\mathbf{M}_{O'y} = \mathbf{u}_y w \left\{\frac{\rho gh}{30}[5h^2 - 4a(5b - 2a)] - \frac{p_a b^2}{2}\right\}$$

$$= (87.9 \times 10^6\ \text{lbf·ft})\ \mathbf{u}_y$$

This indicates the effect of the water and air pressure tending to overturn the dam. This is counteracted by the moment of the gravitational forces on the concrete, and by any capability of the ground attachment at the base and ends for providing additional moments.

Example

The triangular gate shown in Figure 6-8 is hinged about the axis $O\ O'$. To determine the force needed to hold it closed, we calculate the resultant moment about the axis OO' of the water pressure acting on the gate.

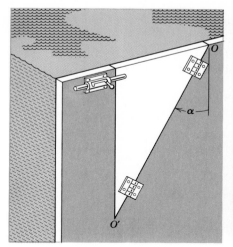 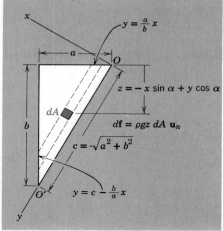

Figure 6-8

Referring to the figure, we see that the moment about this axis (the y component of resultant moment) is given by

$$M_{Oy} = -\int x \, df$$
$$= -\int x \, \rho g(-x \sin \alpha + y \cos \alpha)dA$$
$$= \rho g \sin \alpha \, I_{yy} + \rho g \cos \alpha \, I_{yx}$$

where

$$I_{yy} = \int x^2 \, dA$$
$$= \int_0^{ab/c} \int_{ax/b}^{c-(bx/a)} x^2 \, dy \, dx$$
$$= \int_0^{ab/c} x^2 \left(c - \frac{bx}{a} - \frac{ax}{b} \right) dx$$
$$= \frac{a^3 b^3}{12c^2}$$

and

$$I_{yx} = -\int yx \, dA$$
$$= -\int_0^{ab/c} \int_{ax/b}^{c-(bx/a)} xy \, dy \, dx$$
$$= -\frac{(b^2 + 3a^2)a^2 b^2}{24 \, c^2}$$

Substitution of the values of the integrals into the above results in

$$M_{Oy} = -\frac{\rho g \, a^2 b^3}{24 \, c}$$

Values of integrals like those computed above are available in tabulated summaries like those given in Appendix B of this volume. Certain
general relationships, indicating how to use and extend information
found in the tables, are pointed out in the next section.

Second Moments of Area.   Consider the flat surface area shown in Figure 6-9, acted on by pressure that varies linearly with distance from the

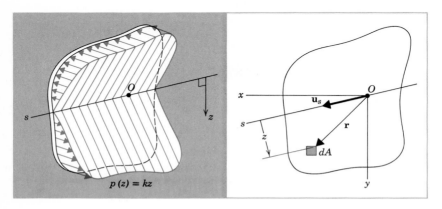

Figure 6-9

axis Os. The resultant moment about O is a vector in the plane of the surface and is related to the pressure gradient through integrals such as
those evaluated above. As is shown in the preceding example, the calculation of moment resultants for this situation arises in the analysis of submerged bodies; another application is in the analysis of bending moments
in elastic beams, a subject outside the scope of this book.

Observe from Figure 6-9 that the distance from the axis Os is given
by

$$z = r \sin \angle_{\mathbf{u}_s}^{\mathbf{r}} = (\mathbf{u}_s \times \mathbf{r}) \cdot \mathbf{u}_n$$

where $\mathbf{u}_n$ is directed outward and perpendicular to the plane of the plate.
The force acting on the area element dA can therefore be expressed as

$$d\mathbf{f} = p \, dA \, \mathbf{u}_n = kz \, \mathbf{u}_n \, dA = k \, \mathbf{u}_s \times \mathbf{r} \, dA$$

and the moment about O can then be expressed as

$$\mathbf{M}_O = \int \mathbf{r} \times d\mathbf{f}$$
$$= k \int \mathbf{r} \times (\mathbf{u}_s \times \mathbf{r}) \, dA$$
$$= k \int [(\mathbf{r} \cdot \mathbf{r})\mathbf{u}_s - (\mathbf{u}_s \cdot \mathbf{r})\mathbf{r}] \, dA$$

Now with the vectors in this equation resolved in the directions of x and y,

$$\mathbf{r} = x\,\mathbf{u}_x + y\mathbf{u}_y$$
$$\mathbf{u}_s = s_x\mathbf{u}_x + s_y\mathbf{u}_y$$

the above can be rearranged to give the x and y components of the moment:

$$\mathbf{M}_O = [k\,I_{xx}s_x + k\,I_{xy}s_y]\,\mathbf{u}_x$$
$$+ [k\,I_{yx}s_x + k\,I_{yy}s_y]\,\mathbf{u}_y \tag{6-6}$$

in which

$$I_{xx} = \int y^2\,dA \qquad I_{yy} = \int x^2\,dA$$
$$I_{xy} = I_{yx} = -\int xy\,dA \tag{6-7}$$

These integrals are called the *second moments of area*,[*] their values depending on the size and shape of the area, and the locations of the axes. We now give some relationships that are useful for extending the information found in the tables in Appendix B.

Parallel axis shift. Let the axes X and Y emanate from the centroid C of a plane area, and let C be located with respect to a parallel pair of axes

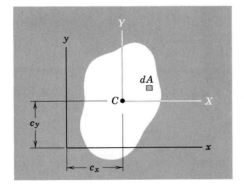

x and y by the coordinates (c_x, c_y). Then the second moments of the area with respect to the two parallel sets of axes are related as follows:

$$I_{xx} = \int y^2\,dA$$
$$= \int (Y + c_y)^2\,dA$$
$$= \int Y^2\,dA + 2c_y\int Y\,dA + c_y^2\int dA$$

---

[*] I_{ii} is also commonly called the *moment of inertia* of the area about the axis i, and I_{xy} the *product of intertia* of the area, about the axes x and y. These terms derive from mathematically analogous integrals that describe inertial resistance to moments in the study of rigid body dynamics. A definition of product of intertia is sometimes made without the minus sign.

Because C is the centroid, $\int Y\,dA = 0$. With the second moment $\int Y^2\,dA$ denoted by I_{XX}, the above may be written as

$$I_{xx} = I_{XX} + c_y{}^2 A \qquad (6\text{-}8a)$$

Similar manipulations lead to

$$I_{yy} = I_{YY} + c_x{}^2 A \qquad (6\text{-}8b)$$

$$I_{xy} = I_{XY} - c_x c_y A \qquad (6\text{-}8c)$$

Thus each of the second moments may be computed by adding, to the corresponding moment with respect to parallel centroidal axes, a term representing the second moment the area would have if it were concentrated at the centroid.

Example

Compute the second moments with respect to x and y of the rectangle in Figure 6-10.

Figure 6-10

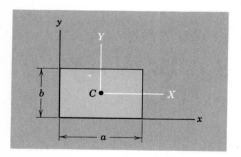

From Appendix B we find that

$$I_{xx} = \frac{ab^3}{3}$$

$$I_{YY} = \frac{ba^3}{12}$$

and from symmetry,

$$I_{XY} = 0$$

Then, from (6-8),

$$I_{yy} = \frac{ba^3}{12} + \left(\frac{a}{2}\right)^2 \quad (ab)$$

$$= \frac{ba^3}{3}$$

$$I_{xy} = 0 - \left(\frac{a}{2}\right)\left(\frac{b}{2}\right)$$

$$= -\frac{a^2 b^2}{4}$$

Rotation of axes. Let the coordinate axes $\bar{x}$ and $\bar{y}$ be rotated counterclockwise through the angle θ, with respect to the axes x and y. Then the coordinates of the element of area dA with respect to the two sets of axes are related as follows:

$$\bar{x} = x \cos \theta + y \sin \theta$$
$$\bar{y} = -x \sin \theta + y \cos \theta$$

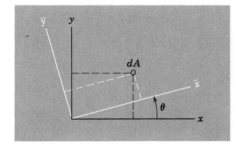

These formulas then can be used to obtain the relationships among the second moments of area with respect to the two sets of axes:

$$\begin{aligned}
\bar{I}_{xx} &= \int \bar{y}^2 \, dA \\
&= \int (-x \sin \theta + y \cos \theta)^2 \, dA \\
&= \cos^2 \theta \int y^2 \, dA - 2 \sin \theta \cos \theta \int xy \, dA + \sin^2 \theta \int x^2 \, dA \\
&= \cos^2 \theta \, I_{xx} + 2 \cos \theta \sin \theta \, I_{xy} + \sin^2 \theta \, I_{yy}
\end{aligned}$$

Similar manipulations lead to

$$\bar{I}_{yy} = \sin^2 \theta \, I_{xx} - 2 \cos \theta \sin \theta \, I_{xy} + \cos^2 \theta \, I_{yy}$$
$$\bar{I}_{xy} = \cos \theta \sin \theta \, (I_{yy} - I_{xx}) + (\cos^2 \theta - \sin^2 \theta) I_{xy}$$

Introducing the trigonometric identities

$$\cos^2 \theta = \frac{1}{2} (1 + \cos 2\theta)$$

$$\sin^2 \theta = \frac{1}{2} (1 - \cos 2\theta)$$

$$\cos \theta \sin \theta = \frac{1}{2} \sin 2\theta$$

into the above results in

$$\bar{I}_{xx} = \frac{1}{2} (I_{xx} + I_{yy}) + \frac{1}{2} (I_{xx} - I_{yy}) \cos 2\theta + I_{xy} \sin 2\theta \qquad (6\text{-}9a)$$

$$\bar{I}_{yy} = \frac{1}{2} (I_{xx} + I_{yy}) - \frac{1}{2} (I_{xx} - I_{yy}) \cos 2\theta - I_{xy} \sin 2\theta \qquad (6\text{-}9b)$$

$$\bar{I}_{xy} = \frac{1}{2} (I_{yy} - I_{xx}) \sin 2\theta + I_{xy} \cos 2\theta \qquad (6\text{-}9c)$$

In the paragraph that follows we describe a useful geometric interpretation of these formulas.

Mohr's circle. Otto Mohr (1835-1918), a German structural engineer, pointed out that Equations 6-9 can be represented by the diagram shown in Figure 6-11. This *Mohr's circle* has the advantages that some of the basic

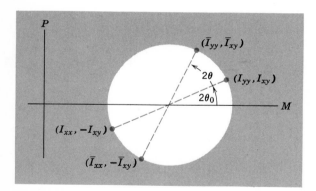

Figure 6-11

properties hidden in the equivalent Equations 6-9 become more readily apparent, and most people find the diagram easier to remember than the equations.

Each point (M,P) on the circle represents a pair of values of second moments for a particular orientation of axes. The abscissa M is equal to the moment of inertia about one axis, and the ordinate P is equal to the product of inertia with respect to that same axis and the axis oriented 90° clockwise from it.

Given values of the second moments with respect to a particular set of axes, the circle may be constructed as follows. A point is located with abscissa equal to I_{yy} and ordinate equal to I_{xy}. Next, another point is located with abscissa equal to I_{xx} and ordinate equal to $-I_{xy}$. (The product of inertia with respect to x and the axis 90° clockwise from x is equal to $-I_{xy}$.) These two points are diametrically opposite on the circle.

Now, any other point on the circle, reached by a rotation of 2θ from any established point corresponds to the second moments with respect to axes rotated θ from those corresponding to the established point on the circle. The rotation in the plane of the Mohr's circle is in the same direction as the rotation of axes in the plane of A.

Example

Construct the Mohr's circle for second moments of area with respect to various sets of axes passing through the corner O, of the rectangle shown in Figure 6-12.

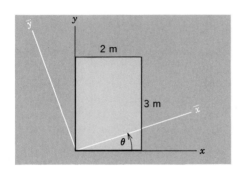

Figure 6-12

From the preceding example, we have

$$I_{xx} = \frac{1}{3} (2 \text{ m})(3 \text{ m})^3 = 18 \text{ m}^4$$

$$I_{yy} = \frac{1}{3} (3 \text{ m})(2 \text{ m})^3 = 8 \text{ m}^4$$

$$I_{xy} = -\frac{1}{4} (3 \text{ m})^2(2 \text{ m})^2 = -9 \text{ m}^4$$

These are used to establish the points $(8, -9)$ and $(18,9)$ as shown in Figure 6-13. The center of the circle is then at the intersection of the M axis and the line connecting these two points. The second moments $\bar{I}_{xx}$ and $\bar{I}_{xy}$, corresponding to the $(\bar{x},\bar{y})$ axes shown in Figure 6-12, are the coordinates of the point so labeled in Figure 6-13. It is evident from the circle that the maximum moment of inertia is equal to 23.3 m⁴ about an axis rotated

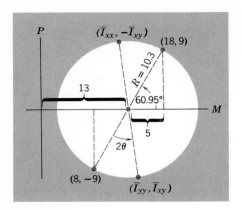

Figure 6-13

$60.95°/2 = 30.47°$ clockwise from x. The minimum moment of inertia is equal to 2.7 m^4, about an axis oriented $90°$ from that of maximum moment of inertia. Furthermore, the product of inertia with respect to these two axes is zero.

To verify the correspondence between Figure 6-11 and Equations 6-9, let the radius of the circle be denoted by R, and the angle from the M axis to (I_{yy}, I_{xy}) by $2\theta_0$, as shown in the figure. From the construction of the circle, the center is established at $M = \frac{1}{2}(I_{xx} + I_{yy})$. Also, from the figure, we have

$$\bar{I}_{yy} = \frac{1}{2}(I_{xx} + I_{yy}) + R \cos(2\theta_0 + 2\theta)$$

$$= \frac{1}{2}(I_{xx} + I_{yy}) + (R \cos 2\theta_0) \cos 2\theta - (R \sin 2\theta_0) \sin 2\theta$$

and

$$R \cos 2\theta_0 = \frac{1}{2}(I_{yy} - I_{xx})$$

$$R \sin 2\theta_0 = I_{xy}$$

Taken together, these relationships from the figure imply that

$$\bar{I}_{yy} = \frac{1}{2}(I_{xx} + I_{yy}) + \frac{1}{2}(I_{yy} - I_{xx}) \cos 2\theta - I_{xy} \sin 2\theta$$

which agrees with Equation 6-9b. In a similar manner, Equations 6-9a and 6-9c can be established from the figure.

Problems

6-8   Determine c, the correction that must be applied to the scale reading to obtain the water surface elevation.

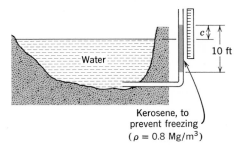

6-9   The gate G is hinged along the top. What force must be applied at the bottom of the gate to prevent its opening?

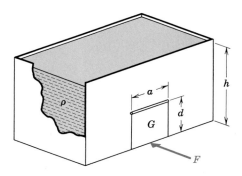

6-10   Locate the line of action of the horizontal component of force on the dam face due to water pressure.

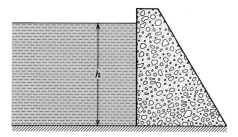

6-11 What is the density of fluid x?

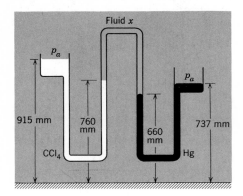

6-12 Determine the air pressure inside the chamber.

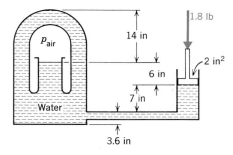

6-13 A hollow sphere, made of steel with a density of 7.8 Mg/m³, has an outside diameter of 600 mm and a wall thickness of 50 mm. How much water will it displace as it floats on the surface?

6-14 The width of the tank (in the direction perpendicular to the plane of the sketch) is w. Determine the tension in the line AB.

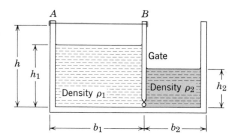

6-15 At what angle θ will the heavy gate be in equilibrium without any support at the top? If it were placed in that position and support removed, would you be willing to stand on the downstream side? Explain carefully.

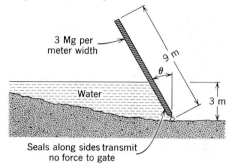

3 Mg per meter width

9 m

θ

Water

3 m

Seals along sides transmit no force to gate

6-16 Consider a free-body consisting of the dam in Figure 6-6 together with the portion of water shown in Figure 6-7. Use equations of equilibrium for this composite body to compute the reaction components at the base of the dam.

6-17 Locate the line of action of the vertical reaction at the bottom of the dam in Figure 6-6, assuming there is no support along the sides of the dam.

6-18 Show that if a plane area has an axis of symmetry, the second moment I_{xy} with respect to that axis and a perpendicular axis is zero.

6-19 Use the geometry of Figure 6-11 to establish consistency with Equations 6-9a and 6-9c.

6-20 Determine the orientation of the axes passing through the corner of the triangle, such that $I_{x_1 x_2} = 0$.

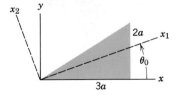

x_2 y

$2a$ x_1

θ_0

x

$3a$

6-21 Determine the orientation of the axes passing through the centroid of the triangle, such that $I_{x_1 x_2} = 0$.

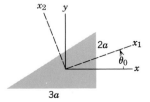

x_2 y

$2a$ x_1

θ_0 x

$3a$

6-22   Find the orientation of the axes passing through the corner of the quarter ellipse, such that $I_{x_1 x_2} = 0$.

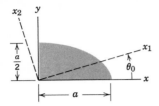

6-23   Determine the second moments of the sector of the circle with respect to the axes shown.

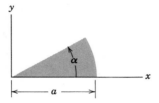

6-3

BENDING MEMBERS.   We learned in Section 5-2 that loads lateral to the axis of a stiff rod will induce internal shear forces and bending moments at various sections of the rod. We also learned that to evaluate them at a particular section, a free-body of the portion of the rod on either side of the section can be subjected to equilibrium analysis.

Here, we are concerned with the manner in which the shear force and bending moment vary from section to section, the usual goal of this analysis being the determination of the maximum bending moment. We describe an approach slightly different from that of Section 5-2, as it sometimes produces the result more easily.

Differential Equations of Equilibrium.   Consider a beam loaded as is shown in Figure 6-14. The loading w is specified as force per unit length of the beam (SI units: newtons per meter). We use the distance x to specify a particular section along the beam and recognize that in general the loading w, the shear force V, and the bending moment M are all functions of x.

Analysis will incorporate the sign convention indicated in the figure: x increases toward the right; positive values of w indicate a downward acting loading; positive values of V indicate a downward acting force on

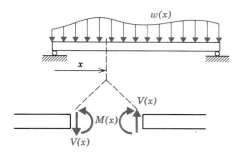

Figure 6-14

the portion of the beam to the left of the section and an upward acting force on the portion of the beam to the right of the section; positive values of M indicate a counterclockwise acting couple on the portion of the beam to the left of the section and a clockwise acting couple on the portion of the beam to the right of the section.

A short element of the beam is shown in Figure 6-15. Vertical force

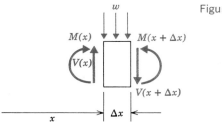

Figure 6-15

equilibrium requires that

$$V(x + \Delta x) - V(x) = -w_{av}\,\Delta x$$

where w_{av} is the average load intensity over the element. After division by Δx this equation becomes, in the limit at $\Delta x \to 0$,

$$\frac{dV}{dx} = -w(x) \qquad (6\text{-}10)$$

Moment equilibrium of the element requires that

$$M(x + \Delta x) - M(x) = V(x)\,\Delta x + (w_{av}\Delta x)\xi\,\Delta x$$

where $\xi\,\Delta x$ is the distance from the single force equivalent of the distributed load, to the right-hand end of the element; we note that $0 < \xi < 1$.

After division by Δx this equation becomes, in the limit as $\Delta x \to 0$,

$$\boxed{\frac{dM}{dx} = V(x)}$$

(6-11)

From the equations of equilibrium (6-10) and (6-11), we see that the shear force $V(x)$ can be obtained by integration of the load intensity $w(x)$, and the bending moment $M(x)$ obtained in turn by integration of the shear force $V(x)$.

Example

Determine the distribution of shear force and bending moment in the cantilever with uniformly distributed load, as depicted in Figure 6-16a.

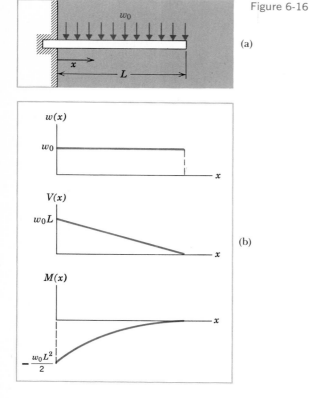

Figure 6-16

(a)

(b)

The load intensity is constant with respect to x:

$$w(x) = w_0$$

Then from (6-10),

$$V(x) = -\int w_0 \, dx = -w_0 x + C_1$$

The constant of integration can be evaluated by means of the boundary condition at the right-hand end; we note that since this end is free of forces

$$V(L) = 0$$

$$-w_0 L + C_1 = 0$$

$$C_1 = w_0 L$$

With this value substituted above, the distribution of shear force is given by

$$V(x) = w_0(L - x)$$

Next, using this result in (6-11), we have

$$M(x) = \int w_0(L - x) \, dx = w_0 \left(Lx - \frac{1}{2} x^2 \right) + C_2$$

The constant of integration arising here can be evaluated by means of the other boundary condition at the right-hand end, where the bending moment must vanish:

$$M(L) = 0$$

$$\frac{1}{2} w_0 L^2 + C_2 = 0$$

$$C_2 = -\frac{1}{2} w_0 L^2$$

With this value substituted above, the distribution of bending moment is given by

$$M(x) = -\frac{1}{2} w_0 (L - x)^2$$

Curves showing the variations of bending moment and shear are shown in Figure 6-16b.

Graphical Interpretations. It is often easier to construct the shear and moment diagrams directly, by means of graphical interpretations of Equations 6-10 and 6-11, than through detailed analytical integration as is shown above.

Recall that on integration of a function that is represented by a given curve, the integrated function exhibits a difference in its values between x_1 and x_2 equal to the area under the given curve between x_1 and x_2. Applying this idea to the first integration in the above example, we see that, between $x = 0$ and some value greater than zero, the value of the shearing force must change by an amount equal to the corresponding area under the loading curve. Because of the minus sign in Equation 6-10 this change must be negative, representing a *decrease* in V. This, together with the observation that V must vanish at $x = L$, permits the drawing of the shear diagram and evaluation of the maximum shear force without resorting to equations. Application of this idea to the next integration gives the bending moment diagram.

Another interpretation is often helpful for determining the forms of the shear and bending moment diagrams, and in checking them for errors. The slope of the bending moment curve is at every point equal to the value of the shearing force at that point. Likewise, the slope of the shearing force curve is at every point equal to the negative of the value of the load intensity at that point. These interpretations follow from the differential forms (6-10) and (6-11).

Concentrated Forces and Couples. A concentrated force is our idealization for a very high intensity of loading over a very short interval of the beam. The magnitude of the force is equal to the area under the short portion of the load diagram over which the high-intensity load acts. In integrating to construct the shear diagram, we must show a decrease in the value of V equal to this area as the integration passes through this region. Thus, consistent with the idealization of a concentrated force, the shear

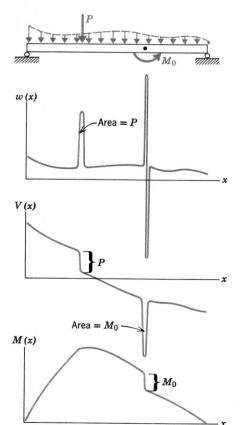

Figure 6-17

diagram exhibits a discontinuity equal to the magnitude of the force, as is illustrated in Figure 6-17. To verify this fact from another point of view, you will find it helpful to consider the equilibrium of a short element of the beam on which the force acts.

An externally applied, concentrated couple is also shown in Figure 6-17. This results in a discontinuity in the bending moment curve, as can be verified by considering the equilibrium of a short element of beam to which the couple is applied.

Example

Determine the distribution of shearing force and bending moment along the beam shown in Figure 6-18.

The values of the support reactions may be determined by con-

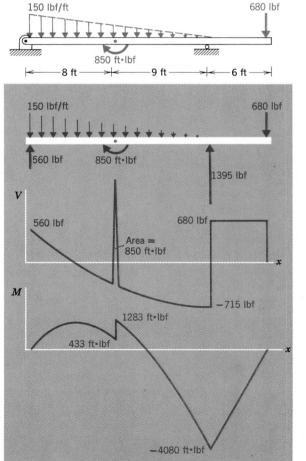

150 lbf/ft   680 lbf Figure 6-18

850 ft·lbf

|← 8 ft →|← 9 ft →|← 6 ft →|

150 lbf/ft   680 lbf

560 lbf 850 ft·lbf

1395 lbf

V

560 lbf   680 lbf

Area =
850 ft·lbf

x

M

−715 lbf

1283 ft·lbf

433 ft·lbf

x

−4080 ft·lbf

sidering equilibrium of the entire beam and are shown on the free-body diagram in the figure.

The reaction at the left gives the value of the shearing force at $x = 0$ as

$$V(0) = 560 \text{ lb}$$

The shape of the shearing force diagram in the region $0 < x < 17$ ft can be deduced from the fact that the slope of this curve must be equal to the negative of the value of $w(x)$ at each point. The equation for this curve may be obtained by integrating Equation 6-10:

$$V = -\int (150 \text{ lb/ft}) \left(1 - \frac{x}{17 \text{ ft}}\right) dx \qquad 0 < x < 17 \text{ ft}$$

$$= 560 \text{ lb} - (150 \text{ lb/ft}) \left(x - \frac{x^2}{34 \text{ ft}}\right) \qquad 0 < x < 17 \text{ ft}$$

This expression does not include an evaluation of the local, high-intensity shearing force that arises from the 850-ft·lbf couple. By considering the equilibrium of a short element on which this couple acts, we find that through this local region the bending moment must increase 850 ft·lbf, while the net change in the shearing force is negligible.

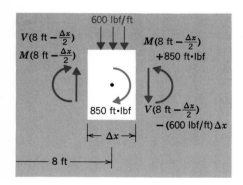

The value of the shearing force just to the left of the right-hand support may be determined by substitution of $x = 17$ ft into the above expression, or by subtracting the area under the loading curve from the value $V(0) = 560$ lbf. The result is

$$V(17 \text{ ft} -) = -715 \text{ lbf}$$

The notation (17 ft –) is to indicate the value of $V(x)$ at a point just to the left of the support.

At this support the 1395-lb reaction induces an increase in the shearing force, as may be seen from a free-body diagram of a short element on which this force acts.

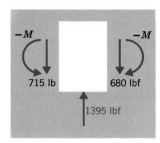

In the region 17 ft $< x <$ 23 ft, the load, and hence the slope of the shearing force curve, is zero. At $x = 23$ ft, the concentrated force induces a change in the shearing force of -680 lbf. Observe that this gives the correct value of shearing force for the region $x > 23$ ft.

The pin support at $x = 0$ implies that the bending moment must be zero at that point. Starting with this initial value, integration of $V(x)$ will give the bending moment at each section. In the region $0 < x < 8$ ft, the result is

$$M(x) = \int \left[560 \text{ lbf} - (150 \text{ lbf/ft}) \left(x - \frac{x^2}{34 \text{ ft}} \right) \right] dx \qquad 0 < x < 8 \text{ ft}$$

$$= (560 \text{ lbf})x - (150 \text{ lbf/ft}) \left(\frac{x^2}{2} - \frac{x^3}{102 \text{ ft}} \right) \qquad 0 < x < 8 \text{ ft}$$

This function reaches a maximum at $x = 4.27$ ft, where the value is $M(4.27) = 1138$ ft·lbf. At $x = 8$ ft, the expression yields

$$M(8 \text{ ft } -) = 433 \text{ ft} \cdot \text{lbf}$$

The concentrated couple applied at $x = 8$ ft induces an increase in the bending moment, as we point out above, so that the bending moment just to the right of that point has the value

$$M(8 \text{ ft } +) = 433 \text{ ft·lbf} + 850 \text{ ft·lbf}$$
$$= 1283 \text{ ft·lbf}$$

In the region 8 ft $< x <$ 17 ft, the bending moment is then given by

$$M(x) = 433 \text{ ft·lbf} + (560 \text{ lbf})x - (150 \text{ lbf/ft}) \left(\frac{x^2}{2} - \frac{x^3}{102 \text{ ft}} \right)$$

Substitution of $x = 17$ ft into this expression yields the value of the bending moment at the right-hand support:

$$M(17 \text{ ft}) = -4080 \text{ ft} \cdot \text{lbf}$$

In the region 17 ft $< x <$ 23 ft, the integration of V yields the straight line with slope of 680 lbf. Observe that the value implied by this curve at $x = 23$ ft agrees with the value we can determine from a free-body of the portion of the beam to the right of $x = 23$ ft.

Problems

6-24 Show a force system different from that acting on the beam, but that is equivalent in the sense discussed in Sections 2-5 and 3-5. Draw shear and bending moment diagrams for the given force system and the equivalent force system.

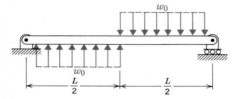

For Problems 6-25 through 6-35, draw shear and bending moment diagrams, labeling all values at important points.

6-25

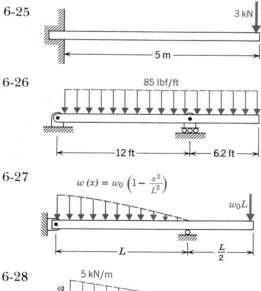

6-26

6-27

6-28

6-29

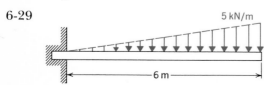
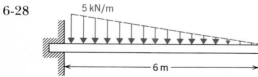

6-30

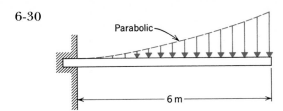

Same total load as in Problems 6-28, 6-92.

6-31

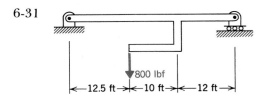

6-32

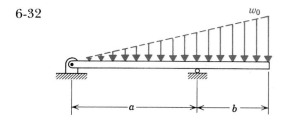

6-33

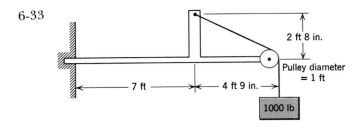

6-34

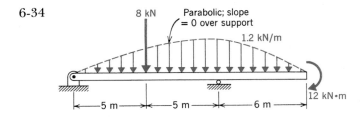

6-35

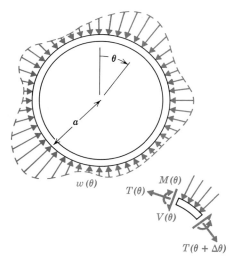

6-36 Radial loads $w(\theta)$ will induce internal forces V (shear), T (hoop tension), and M (bending moment) in the circular ring, as is indicated on the element. Derive the differential relationships connecting $w(\theta)$, $V(\theta)$, $T(\theta)$, and $M(\theta)$.

6-37 Loads $w(\theta)$ perpendicular to the plane of the circular rod will induce internal forces V (shear), M_b (bending moment), and M_t (twisting moment). Derive the differential relationships connecting $w(\theta)$, $V(\theta)$, $M_b(\theta)$, and $M_t(\theta)$.

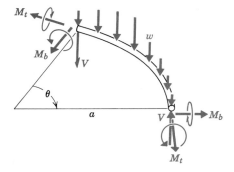

6-38   Evaluate $V(\theta)$, $M_b(\theta)$, and $M_t(\theta)$.

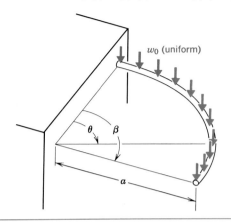

w_0 (uniform)

θ β

a

6–4

FLEXIBLE LINES.   Cables and chains are often subjected to lateral forces which induce internal tension. Because of their inherent flexibility, the bending moments that these members can transmit are usually negligible, and from the equilibrium analysis of beams we know that this implies shearing forces must also be negligible. That is, the resultant internal force between elements must be in the direction tangent to the axis of the cable or chain.

The flexibility of the line adds a feature to equilibrium problems that has been absent from the other examples we have considered thus far. In addition to unknown forces, the configuration of these bodies is not normally known á priori; therefore, an additional unknown quantity enters such problems. In this section we examine the configuration and tension forces induced in flexibile lines by distributed forces that are all parallel.

Consider the suspended line shown in Figure 6-19, subjected to a distributed vertical load intensity w (force per unit of horizontal distance). With the tension denoted by S, and the angle from the horizontal to the tangent to the line by α, horizontal equilibrium of a portion of the cable requires that

$$S \cos \alpha = S_0 \text{ (constant)} \tag{6-12}$$

That is, the horizontal component of tension is constant. This implies that *the tension is maximum where the configuration of the line is steepest.*

Vertical equilibrium of the short element of line requires that

$$(S \sin \alpha)_{x+\Delta x} - (S \sin \alpha)_x = w_{av} \Delta x$$

where w_{av} is the average load intensity within the interval Δx. After

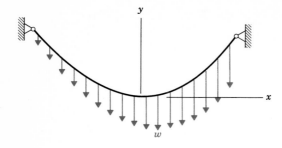

Figure 6-19

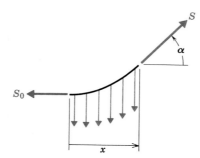

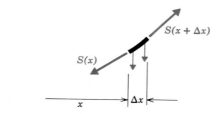

division by Δx, this equation becomes, in the limit as $\Delta x \to 0$,

$$\frac{d}{dx} (S \sin \alpha) = w(x)$$

Or, after using (6-12) to eliminate the tension resultant,

$$\frac{d}{dx} (S_0 \tan \alpha) = w(x)$$

But $\tan \alpha = dy/dx$, so that

$$S_0 \frac{d^2y}{dx^2} = w(x) \qquad (6\text{-}13)$$

Given the load distribution and support points, this equation determines the configuration by specifying $y(x)$. Once this is determined, the tension can be evaluated by means of Equation 6-12:

$$S = S_0 \sec \alpha$$
$$= S_0 \sqrt{1 + \tan^2 \alpha}$$
$$= S_0 \sqrt{1 + (dy/dx)^2} \tag{6-14}$$

We now examine two commonly encountered types of load distribution. Under more complicated load distributions, the differential equation 6-13 may require numerical integration for its solution.

Horizontally Uniform Loading. Suspension bridge decks are normally suspended from the main cables by closely spaced vertical lines. These lines transmit to each main cable a load that is of nearly uniform intensity in the horizontal direction. In this case, we have

$$w(x) = w_0 \qquad \text{(constant)}$$

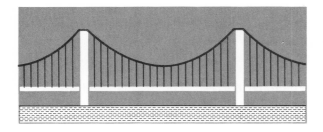

and Equation 6-13 may be readily integrated to give the parabolic configuration

$$S_0 y(x) = \frac{1}{2} w_0 x^2 + C_1 x + C_2$$

If we select the origin of the x-y coordinates at the apex of the parabola, the constants of integration C_1 and C_2 are both zero, and

$$y = \frac{w_0 x^2}{2 S_0} \tag{6-15}$$

Example

The deck of a suspension bridge weighs 42 tons per foot of length. The two main cables are suspended between towers 2800 ft apart. The top of each tower is 320 ft above the lowest point of a main cable. What is the maximum tension in each main cable?

The maximum tension will occur at the attachment with the tower, where $x = 1400$ ft. The load intensity on each of the two cables is

$$w_0 = \frac{1}{2} \,(42 \text{ tons/ft})$$
$$= 21 \text{ tons/ft}$$

Using the coordinates of the attachment point in Equation 6-15, the horizontal component of tension is computed as

$$S_0 = \frac{(21 \text{ tons/ft})(1400 \text{ ft})^2}{2\,(320 \text{ ft})}$$
$$= 64\ 300 \text{ tons}$$

The tension at the tower may now be determined from Equation 6-14:

$$S = S_0 \,\sqrt{1 + (dy/dx)^2}$$
$$= \sqrt{S_0{}^2 + (w_0 x)^2}$$
$$= \sqrt{(64\ 300 \text{ tons})^2 + [(21 \text{ tons/ft})(1400 \text{ ft})]^2}$$
$$= 70\ 700 \text{ tons}$$

Uniform Line Under Gravitational Forces. The gravitational forces exerted on the suspended line itself are of varying intensity with respect to horizontal distance, because of the varying slope of the line. For a uniform cable with density ρ and cross sectional area A, this load intensity is given by

$$w_g(x) = \rho A g \sec \alpha$$

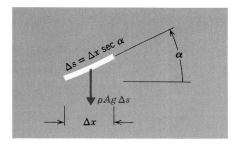

We see from the equation preceding (6-13) that when these are the only significant forces, the equilibrium curve must satisfy

$$S_0 \,\frac{d}{dx} \tan \alpha = \rho A g \sec \alpha$$

With the abbrevations

$$u = \tan \alpha = \frac{dy}{dx}$$

$$a = \frac{S_0}{\rho A g} \qquad (6\text{-}16)$$

this equation can be rewritten as

$$\frac{du}{\sqrt{1 + u^2}} = \frac{dx}{a}$$

Integration results in

$$\sinh^{-1} u = \frac{x}{a} + C_1$$

Or,

$$u = \sinh \left(\frac{x}{a} + C_1 \right)$$

Another integration gives

$$y = \int u \, dx$$

$$= a \cosh \left(\frac{x}{a} + C_1 \right) + C_2$$

Now, if we select the origin of the x-y coordinates at the apex of the curve, where $y = \frac{dy}{dx} = 0$

$$C_1 = 0, \qquad C_2 = -a$$

and the equilibrium curve is then given by

$$y = a \left(\cosh \frac{x}{a} - 1 \right) \qquad (6\text{-}17)$$

The curve given by this equation is called the *common catenary*. Several of its properties, important in the analysis of suspended cables, are developed in the following.

Useful relationships of catenaries. The angle α between the tangent to the catenary and the horizontal is given by

$$\sec \alpha = \frac{ds}{dx}$$

$$= \sqrt{1 + \left(\frac{dy}{dx} \right)^2}$$

$$= \sqrt{1 + \sinh^2 \left(\frac{x}{a} \right)}$$

$$= \cosh \left(\frac{x}{a} \right) \qquad (6\text{-}18)$$

Integration then gives the arc length, measured from the apex, as

$$s = \int \left(\frac{ds}{dx}\right) dx$$

$$= \int \cosh \left(\frac{x}{a}\right) dx$$

$$= a \sinh \left(\frac{x}{a}\right) \tag{6-19}$$

But since $\tan \alpha = dy/dx = \sinh x/a$, we also have

$$\tan \alpha = \frac{s}{a} \tag{6-20}$$

which tells us that *the parameter a is equal to the arc length from the apex to the point where $\alpha = \pi/4$.*

The tension at any point is related to the horizontal component S_0 by

$$S = S_0 \sec \alpha$$

and, in view of Equation 6-18, is given by

$$S = S_0 \cosh \left(\frac{x}{a}\right) \tag{6-21}$$

We next study the determination of the catenary size parameter a in terms of the length of line and the positions of the supports. Let the coordinates (x_1, y_1) and (x_2, y_2) locate the end points, as shown in Figure 6-20. The length of line l and support elevation difference h may be

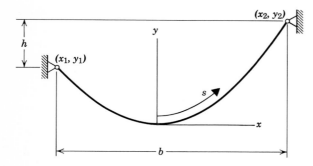

Figure 6-20

expressed by

$$l = s_2 - s_1 = a \sinh \left(\frac{x_2}{a}\right) - a \sinh \left(\frac{x_1}{a}\right)$$

$$h = y_2 - y_1 = a \cosh \left(\frac{x_2}{a}\right) - a \cosh \left(\frac{x_1}{a}\right)$$

Addition and substraction of these two equations lead to

$$e^{x_2/a} - e^{x_1/a} = \frac{(l+h)}{a}$$

$$-e^{-x_2/a} + e^{-x_1/a} = \frac{(l-h)}{a}$$

These may then be solved simultaneously, with the result

$$\frac{x_2}{a} = \log\left[(p+1)\frac{(l+h)}{2a}\right] \tag{6-22a}$$

$$\frac{x_1}{a} = \log\left[(p-1)\frac{(l+h)}{2a}\right] \tag{6-22b}$$

in which

$$p = \sqrt{1 + \frac{4a^2}{l^2 - h^2}} \tag{6-23a}$$

Now, the horizontal distance between the end points may be expressed by

$$\frac{b}{a} = \frac{x_2}{a} - \frac{x_1}{a} = \log\frac{p+1}{p-1}$$

from which

$$p = \frac{e^{b/a} + 1}{e^{b/a} - 1} = ctnh\left(\frac{b}{2a}\right) \tag{6-23b}$$

Finally, elimination of p from Equations 6-23a and 6-23b leads to

$$\frac{b/2a}{\sinh(b/2a)} = \frac{b}{\sqrt{l^2 - h^2}} \tag{6-24}$$

This equation defines the ratio a/b for each value of the parameter $b/\sqrt{l^2 - h^2}$. Viewed in this manner, Equation 6-24 is transcendental; that is, determination of a/b requires a trial-and-error procedure. Table 6-1

Table 6-1

$b/\sqrt{l^2 - h^2}$	a/b	$b/\sqrt{l^2 - h^2}$	a/b
0.00	0.000 00	0.35	0.180 93
0.01	0.068 64	0.40	0.195.87
0.02	0.077 25	0.45	0.211 97
0.03	0.083 48	0.50	0.229 64
0.04	0.088 62	0.55	0.249 39
0.05	0.093 12	0.60	0.271 95
0.10	0.111 11	0.65	0.298 34
0.15	0.125 97	0.70	0.330 15
0.20	0.139 73	0.75	0.370 03
0.25	0.153 20	0.80	0.422 75
0.30	0.166 81	0.85	0.498 28

and the corresponding curve of Figure 6-21 give values satisfying Equation 6-24.

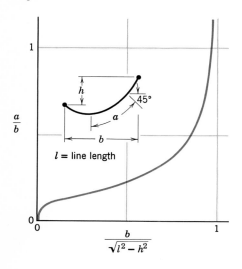

Figure 6-21

By using the first three terms in the power series expansion

$$sinh\ \theta = \theta + \frac{\theta^3}{3!} + \frac{\theta^5}{5!} + \cdot\ \cdot\ \cdot$$

the following approximation can be developed:

$$\frac{a}{b} \approx \left(40\left\{\left[1 + \frac{6}{5}\left(\frac{1-u}{u}\right)\right]^{\frac{1}{2}} - 1\right\}\right)^{-\frac{1}{2}} \tag{6-24a}$$

in which

$$u = \frac{b}{\sqrt{l^2 - h^2}}$$

This approximation is 0.38% low at $u = 0.65$ and becomes increasingly accurate for larger u.

Example

A 1300-ft cable, weighting 6.5 tons, is to be suspended between two supports that are 780 ft apart horizontally and are at an elevation difference of 500 ft. Determine the maximum tension induced by gravity.

The sag parameter in Equation 6-24 has the value

$$\frac{b}{\sqrt{l^2 - h^2}} = \frac{(780)}{\sqrt{(1300)^2 - (500)^2}}$$
$$= 0.650$$

Then from Table 6-1, we obtain

$$\frac{a}{b} = 0.2983$$

from which

$$a = (0.2983)(780 \text{ ft}) = 232.7 \text{ ft}$$

The largest tension occurs at $x = x_2$, which may be determined as follows. From Equation 6-23b,

$$p = \frac{e^{1/0.2983} + 1}{e^{1/0.2983} - 1} = 1.0726$$

Then, from Equation 6-22a

$$\frac{x_2}{a} = \log \left[\frac{(2.0726)(1300 + 500)}{2 (232.7)} \right]$$

$$= 2.081$$

The tension at this point may now be computed by using Equations 6-16 and 6-21.

$$S_2 = (\rho A g) \; a \; \cosh \left(\frac{x_2}{a} \right)$$

$$= \frac{6.5 \text{ tons}}{1300 \text{ ft}} (232.7 \text{ ft}) \cosh 2.081$$

$$= 4.74 \text{ tons}$$

$$= 42.1 \text{ kN}$$

A Problem in cable laying.   Ocean moorings and other flexible line installations have given rise to the following problem: when one of the attachment points is on a solid horizontal floor, an unknown portion of the cable may lie along the floor, and the remainder will form a catenary (in the absence of all but gravitational loading, of course). Given the total length of cable L and the relative positions of the attachment points, in terms of B and h as shown in Figure 6-22, we wish to determine the geometry of the catenary portion so that force analysis can be carried out.

The length of the portion of cable in contact with the floor is equal to the difference between the total length L and the length l of the catenary portion:

$$L - l = B - b$$

Or,

$$l - b = L - B$$

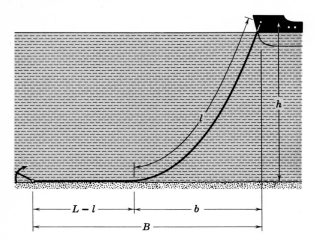

Figure 6-22

Equilibrium of a line element at the point where the cable parts from the floor shows that the catenary is tangent to the floor at that point. Therefore, we can consider the apex of this particular catenary as one of its attachment points; that is, we can use the equations for the general catenary with $x_1 = 0$ and $x_2 = b$. With these values, the equations for l and h, at the bottom of p. 164, then yield

$$\frac{l}{h} = \frac{\sinh (b/a)}{\cosh (b/a) - 1} \tag{6-25}$$

$$\frac{b}{h} = \frac{b/a}{\cosh (b/a) - 1} \tag{6-26}$$

Substitution into the above length relationship then gives

$$\frac{\sinh (b/a) - b/a}{\cosh (b/a) - 1} = \frac{L - B}{h} \tag{6-27}$$

Given L, B, and h, this equation determines the ratio a/b. This done, the values of l/h and b/h can be computed from Equations 6-25 and (6-26). To facilitate solution of the transcendental equation (6-27) for a/b, the curves in Figure 6-23 have been plotted. With a value of $(L - B)/h$ entered on the vertical scale on the right, the value of a/b may be read on the abscissa. Corresponding values of l/h and b/h may then be read by using the other curves.

Approximations can be constructed by using the lead terms in the power series developments of the functions of b/a in Equations 6-25, 6-26, and 6-27. This procedure results in

$$\frac{l}{h} \approx \frac{2h}{3(L - B)} \left[1 + \frac{9}{20} \left(\frac{L - B}{h} \right)^2 \right] \tag{6-25a}$$

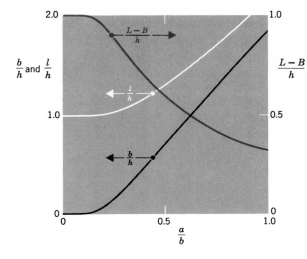

Figure 6-23

$$\frac{b}{h} \approx \frac{2h}{3(L-B)}\left[1 - \frac{21}{20}\left(\frac{L-B}{h}\right)^2\right] \qquad (6\text{-}26a)$$

$$\frac{a}{b} \approx \frac{h}{3(L-B)}\left[1 - \frac{3}{10}\left(\frac{L-B}{h}\right)^2\right] \qquad (6\text{-}27a)$$

At a value of $(L-B)/h = 0.4197$ (where $a/b = 0.75$), the above approximation for l/h is 0.09% low, the approximation for b/h is 0.12% low, and the approximation for a/b is 0.30% high. The accuracies improve with decreasing values of $(L-B)/h$.

Example

One end of an 1870-m cable is anchored on the horizontal sea floor 750 m below the surface. The upper end is attached to a ship at a point 1600 m south of the anchor. The difference between the gravitational and buoyant forces acting on the submerged cable is 131 N/m of cable length. What is the force exerted by the cable on the ship?

The net force intensity is

$$\rho_n Ag = 131 \text{ N/m}$$

and the lengths shown in Figure 6-22 are

$$L = 1870 \text{ m}$$
$$B = 1600 \text{ m}$$
$$h = 750 \text{ m}$$

The ratio appearing in the right-hand side of (6-27) has the value

$$\frac{(L-B)}{h} = \frac{1870-1600}{750} = 0.360$$

Using this value in Figure 6-23, we obtain

$$\frac{l}{h} = 1.96$$

$$\frac{b}{h} = 1.60$$

$$\frac{a}{b} = 0.89$$

Then,

$$l = (1.96)(750 \text{ m}) = 1470 \text{ m}$$
$$b = (1.60)(750 \text{ m}) = 1200 \text{ m}$$
$$a = (0.89)(1200 \text{ m}) = 1068 \text{ m}$$

Now, from (6-16), the horizontal component of tension is

$$S_0 = (131 \text{ N/m})(1068 \text{ m}) = 140 \text{ kN}$$

and from (6-21) the resultant tension at the attachment with the ship is

$$S_2 = (140 \text{ kN}) \cosh \frac{1200}{1068} = 238 \text{ kN}$$

This force acts at an angle α_2 with the horizontal given by

$$\cos \alpha_2 = \frac{140}{238}$$

Or,

$$\alpha_2 = 54°$$

Problems

6-39 The lineal density of the cable in the example on pages 161 to 162 is 9 Mg/m. What is the additional tension at the tower caused by the forces of gravity on the cable itself?

6-40 Evaluate the compressive force in the tower of the suspension bridge discussed in the example on pages 161 to 162.

6-41 The towers of a suspension bridge are 430 m apart, and the points of cable attachment are 50 m above the lowest points on the cable. The maximum tension in each cable is 150 MN. What is the minimum tension in each cable, and what is the deck loading?

6-42   A uniform cable is to be strung between the two points as is shown below. The cable is to be tangent with the horizontal at the lower attachment point. What length cable is required, and what is the maximum tension if the cable weighs 10 lb/ft?

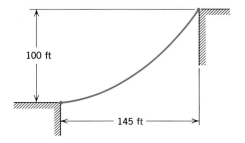

6-43   The 735-kg cable supports the 400-kg load at mid-span. Before the concentrated load was added, the sag in the cable was 30 m. Determine the final sag h_1, and the maximum tension in the cable.

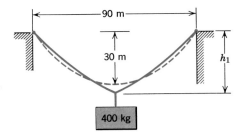

6-44   Show that for a given cable length and material, and for given support positions, the stress (tension divided by cross sectional area) in the cable loaded only by gravitational forces on the cable itself, is independent of the cable diameter.

6-45   What is the largest value the parameter $b/\sqrt{l^2 - h^2}$, appearing in Equation 6-24, can have?

6-46   For given values of B and h, shown in Figure 6-22, what is the shortest length L for which some of the cable will lie along the ocean floor?

6-47   Show that the apex of a common catenary will lie between the support points (x_1, y_1) and (x_2, y_2), provided that

$$\frac{ah}{l^2 - h^2} \leq \frac{1}{2}$$

Taking account of (6-24), this criterion has the form

$$\frac{b}{h} \geq f \frac{b}{(\sqrt{l^2 - h^2}\,)}$$

Plot this criterion and Equation 6-24, using a common abscissa $b/\sqrt{l^2 - h^2}$, and explain how this can be used to solve Problem 6-42 and others with different support positions.

6-48   A 140-ft line having a density of 0.9 Mg/m³ runs from the dock to the anchor as shown. The density of seawater is about 1.03 Mg/m³. Estimate the line tension at the anchor.

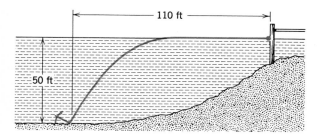

6-49   Derive the approximation (6-24a), and plot the error versus u through the range of values given in Table 6-1.

VIRTUAL WORK

Some mechanical systems, particularly complex mechanisms, are better suited to an approach to equilibrium analysis which is somewhat different from that taken in Chapters 4 through 6. The main advantage of this alternative approach is that it often gives directly an equation which would otherwise require a number of intermediate steps for the elimination of unknown forces from a system of equations. The idea is based on the notion of work, which we will pursue in some detail in the study of dynamics.

7-1

WORK DONE BY A FORCE. The work done by a force $\mathbf{f}$, as the particle on which it acts undergoes a change in its position, is defined as

$$W_{1-2} = \int_{\mathbf{r}_1}^{\mathbf{r}_2} \mathbf{f} \cdot d\mathbf{r} \qquad (7\text{-}1)$$

where $\mathbf{r}$ is a varying position vector that locates the particle relative to some stationary reference point. We observe that the component of force

173

perpendicular to the displacement increment does not contribute to the work increment.

For example, the work done by the force of gravity on a 7.26 kg shotput as it travels from a height of 2.0 m to a height of 4.5 m is com-

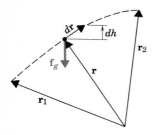

puted as follows. Noting that in this case

$$\mathbf{f}_g \cdot d\mathbf{r} = f_g \, dr \cos \angle \begin{smallmatrix} d\mathbf{r} \\ \mathbf{f}_g \end{smallmatrix}$$
$$= -f_g \, dh$$

we write

$$W_{1-2} = \int_{\mathbf{r}_1}^{\mathbf{r}_2} \mathbf{f}_g \cdot d\mathbf{r}$$

$$= \int_{2.0\,\mathrm{m}}^{4.5\,\mathrm{m}} -(7.26 \text{ kg})(9.81 \text{ N/kg}) \, dh$$

$$= -178 \text{ N·m}$$

The SI unit for work is the newton meter, called the *joule* (J = N·m).

Problems

7-1 Calculate the work done in stretching the spring from its relaxed position to the displacement x. Express the result in terms of the stiffness k, and
(a) the spring extension x;
(b) the force at the extended configuration.

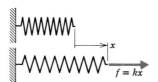

7-2   If the shotput in the example is slowly *lifted* from 2 m to 4.5 m, what is the work done by the upward-directed force necessary to overcome the gravitational force? What is the sum of the works done by the lifting and gravitational forces?

7-3   What is the work done by a lifting force as it slowly raises an object of mass m from the surface of the earth to an altitude of 1000 km? Compare this with an estimate made under the assumption that the magnitude of the force is constant, equal to mg.

7–2

WORK DONE BY FORCES ON A RIGID BODY.   As a rigid body undergoes an incremental change in position, the increment of work done by the forces acting on the body is defined as the sum of the work increments done by the individual forces. Referring to Figure 7-1a, we write this as

$$dW = \sum_i (\mathbf{f}_i \cdot d\mathbf{r}_i)$$

Figure 7-1

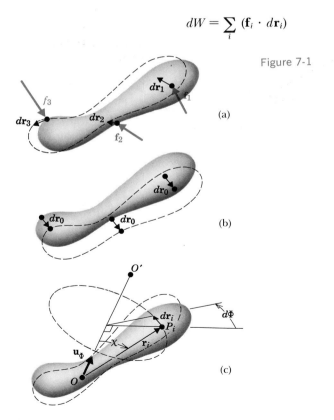

(a)

(b)

(c)

Consider first a simple displacement increment in which every point moves parallel to all other points as depicted in Figure 7-1*b*; that is, no rotation takes place. The displacement of all points will then be equal,

$$d\mathbf{r}_i = d\mathbf{r}_0$$

and the increment of work becomes

$$dW = \sum_i (\mathbf{f}_i \cdot d\mathbf{r}_0)$$

$$= \left(\sum_i \mathbf{f}_i\right) \cdot d\mathbf{r}_0$$

$$= \mathbf{f} \cdot d\mathbf{r}_0$$

in which $\mathbf{f} = \Sigma\mathbf{f}_i$ is the resultant force acting on the body. Therefore, in the absence of rotation, the work done on the rigid body is calculated in the same way as that done on a particle.

Next, consider a small rotation about the axis OO', as depicted in Figure 7-1*c*. We denote the angle of rotation (in radians) by $d\Phi$. Now, observe from the figure that the magnitude of the position change of point P_i may be expressed as

$$dr_i = \overline{OP}_i \sin \chi \, d\Phi$$

But this is also the magnitude of $\mathbf{u}_\Phi \times \mathbf{r}_i \, d\Phi$, where $\mathbf{u}_\Phi$ is a unit vector in the direction of the axis of rotation, and $\mathbf{r}_i$ is a vector from any point on the axis OO' to the point P_i. Finally, if we define the direction of rotation using the right-hand rule in conjunction with the direction of $\mathbf{u}_\Phi$, we can write the displacement increment as

$$d\mathbf{r}_i = (d\Phi \, \mathbf{u}_\Phi) \times \mathbf{r}_i$$

in which the direction and magnitude of the rotation of the body are completely characterized by the vector $d\Phi \, \mathbf{u}_\Phi$. Then the increment of work done by a set of forces acting on the body as it undergoes such an increment of rotation is given by

$$dW = \Sigma(\mathbf{f}_i \cdot d\mathbf{r}_i)$$
$$= \Sigma[\mathbf{f}_i \cdot (d\Phi \, \mathbf{u}_\Phi) \times \mathbf{r}_i]$$
$$= \Sigma[(\mathbf{r}_i \times \mathbf{f}_i) \cdot (d\Phi \, \mathbf{u}_\Phi)]$$
$$= (\Sigma\mathbf{r}_i \times \mathbf{f}_i) \cdot (d\Phi \, \mathbf{u}_\Phi)$$
$$= \mathbf{M}_O \cdot \mathbf{u}_\Phi \, d\Phi$$

in which $\mathbf{M}_O$ is the resultant moment of the forces about O. We note that the component of moment perpendicular to the axis of the rotation increment does not contribute to the work increment.

An *arbitrary* change in position of a rigid body can be expressed as a superposition of a translation equal to the displacement of any point O and a rotation about an axis through O. For a given repositioning, the translation depends on the point O selected, but the rotation $\mathbf{u}_\Phi d\Phi$ is independent of this selection. Far from obvious, these facts will be demonstrated in later chapters on kinematics. The increment of work done on a rigid body as it undergoes an arbitrary increment of change in position is, therefore,

$$dW = \Sigma \mathbf{f}_i \, (d\mathbf{R}_O + d\Phi \, \mathbf{u}_\Phi \times \mathbf{r}_i)$$
$$= \mathbf{f} \cdot d\mathbf{R}_O + \mathbf{M}_O \cdot \mathbf{u}_\Phi \, d\Phi$$

The work done as the rigid body undergoes a repositioning is then

$$\boxed{W_{1-2} = \int_1^2 [\mathbf{f} \cdot d\mathbf{R}_O + \mathbf{M}_O \cdot \mathbf{u}_\Phi \, d\Phi]} \tag{7-2}$$

In general, $\mathbf{f}$, $\mathbf{M}_O$, and $\mathbf{u}_\Phi$ can all vary as the body is moved a finite amount.

Example

An upward force is applied to the end of the wrench handle, 440 mm from the elbow shown in Figure 2-4. Friction in the threads requires that a moment about the pipe axis of 44.0 N·m be applied to loosen it. What will be the work done by the force as the elbow is unscrewed 2°?

The component of moment in the direction of the axis of rotation is 44.0 N·m. Then, assuming this moment does not vary as the 2° rotation takes place, we have

$$W = \mathbf{M}_O \cdot \mathbf{u}_\Phi \int_0^{2°} d\Phi$$
$$= (44.0 \text{ N·m}) \, (2°) \, (\pi \text{ rad}/180°)$$
$$= 1.54 \text{ J}$$

It is instructive to evaluate the applied force and linear travel at the end of the handle as the rotation takes place, and to compute the work using (7-1).

Problems

7-4   Evaluate the work done on the winch handle in raising the car a height of 2 ft. The length of the winch handle is 2 ft and the spool onto which the line is wound is 6 in. in diameter.

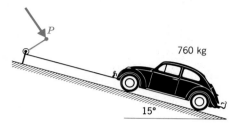

7-5   Evaluate the work done by the force T in raising the load to a height h, for each of the mechanisms of Problem 4.20.

7–3

THE PRINCIPLE OF VIRTUAL WORK.   The concept of work embodied in the above definitions is useful in the study of dynamics, because of its relationship to changes in certain quantities associated with motion. The closely related concept of *virtual work* is useful in both statics and dynamics.

Virtual work is defined as the work that would be done by all the forces acting on a system if the parts within the system were to undergo small increments of displacement. These increments of displacement, called the components of a *virtual displacement* of the system, are arbitrary; this means that we may consider as a virtual displacement any small change in configuration that suits the needs of the problem at hand, whether physically realizable or not.

Presently, we will show the equivalence between the previously used conditions for equilibrium and the *principle of virtual work*, which states: *A mechanical system is in static equilibrium if and only if the virtual work* δW *is zero for an arbitrary virtual displacement.* However, most readers will grasp the general idea more readily after first examining some examples.

Example

The slider-crank mechanism shown in Figure 7-2a has a vertical force P applied to the piston and a couple with moment M applied to the crank. Assuming negligible friction, we want to determine the relationship between P and M for equilibirum.

Consider a virtual displacement of the system depicted in Figure 7-2b, in which the crank and connecting rod do not deform. The upper end of the connecting rod undergoes a small vertical displacement δx, and

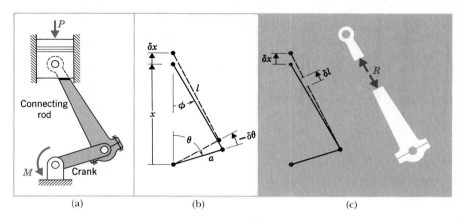

Figure 7-2

the crank undergoes a corresponding increment of angular displacement $-\delta\theta$. The corresponding virtual work of the forces is computed in the same manner as an increment of work on an actual set of displacements; with reference to the equation preceding (7-2), this virtual work may be written as

$$\delta W = -P\,\delta x + M(-\delta\theta)$$

Before we can make use of this, we must determine the geometric relationship between δx and $\delta\theta$. To do this, we begin with two equations from the triangle formed by the crank, the rod, and the vertical:

$$x = a\,\cos\,\theta + l\,\cos\,\phi$$
$$a\,\sin\,\theta = l\,\sin\,\phi$$

Differentiation gives the first-order relationships among the variations in x, θ, and ϕ, as

$$\delta x = -a\,\sin\,\theta\,\delta\theta - l\,\sin\,\phi\,\delta\phi$$
$$a\,\cos\,\theta\,\delta\theta = l\,\cos\,\phi\,\delta\phi$$

Algebraic elimination of $\delta\phi$ from these two equations gives us the variation in x in terms of the variation in θ:

$$\delta x = -a(\sin\,\theta + \tan\,\phi\,\cos\,\theta)\,\delta\theta$$

using this result, we can write the virtual work expression as

$$\delta W = [Pa(\sin\,\theta + \tan\,\phi\,\cos\,\theta) - M]\,\delta\theta$$

Now, according to the principle of virtual work, if the system is in equilibrium, this is zero for an arbitrary value of $\delta\theta$. Therefore, we have

$$M = Pa(\sin \theta + \tan \phi \cos \theta)$$

If desired, a little trigonometry can be used to eliminate ϕ, with the result that

$$M = Pa \sin \theta \left(1 + \frac{(a/l) \cos \theta}{\sqrt{1 - (a^2/l^2) \sin^2 \theta}}\right)$$

Example

Determine the force in the connecting rod in Figure 7-2a.

For this problem, we consider a virtual displacement on which the desired force would do work. A variation in x with no variation in θ or deformation of the crank would require a variation in l, as shown in Figure 7-2c. This can be visualized as a separation somewhere along the rod, and we see that the internal surface force acting on the upper portion would do work equal to $R\ \delta l$. The relationship between δl and δx can be determined, as in the preceding example, by differentiating the equations for the triangle to obtain

$$\delta x = \cos \phi\ \delta l - l \sin \phi\ \delta\phi$$
$$0 = \sin \phi\ \delta l + l \cos \phi\ \delta\phi$$

and algebraically eliminating $\delta\phi$, with the result that

$$\delta l = \cos \phi\ \delta x$$

The virtual work is then

$$\delta W = -P\ \delta x + R\ \delta l$$
$$= (-P + R \cos \phi)\ \delta x$$

and, because this must be zero for arbitrary δx,

$$R = P \sec \phi$$

The geometric analysis of the virtual displacements can usually be carried out more directly by sketching the small displacements, rather than resorting to formal differentiation as was done in the preceding examples. For example, the change in length of the rod in the last example can be determined as follows. Let the rod pivot a small amount about the crankpin, moving its upper end upward and to the right. Then let the rod lengthen until its upper end is at a point directly above its original position. The small triangle formed by the vertical and the path traced out by

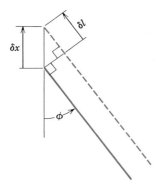

the upper end shows at once that

$$\delta l = \delta x \cos \phi$$

Example

Figure 7-3a is a schematic of a platform balance scale. Determine the relationship among the dimensions so that the force ratio R/f_g is independent

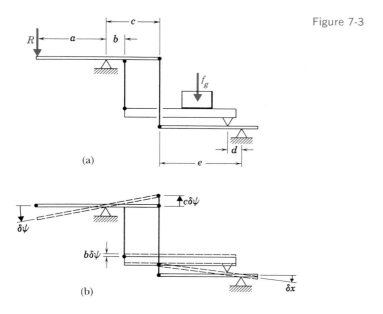

Figure 7-3

of the position of the body on the platform. Also, determine the force ratio in terms of the dimensions.

Consider a virtual displacement of the system shown in Figure 7-3b, in which none of the members deforms. The virtual work principle will have the form

$$R\,\delta_R - f_g\,\delta_g = 0$$

in which δ_R and δ_g are the displacements of the points of application of the two forces. But if the value of R satisfying this is to be independent of the position of f_g along the platform, δ_g must also be independent of this position, which implies that during the virtual displacement the platform must remain level. In analyzing the geometry of the various components of the displacement, we observe that the vertical displacement δy of a point located a distance r from the pivot of a horizontal rod is related to the small angle of rotation $\delta\theta$ by

$$\delta y = r\,\delta\theta$$

Thus the right-hand end of the upper rod undergoes a displacement equal to $c\delta\psi$. Since this is equal to the displacement of the left-hand end of the lower rod,

$$e\,\delta\chi = c\,\delta\psi$$

We can express the requirement that the platform remain level as

$$b\,\delta\psi = d\,\delta\chi$$
$$= d\left(c\,\frac{\delta\psi}{e}\right)$$

which yields, as the condition for independence of load position,

$$\frac{b}{d} = \frac{c}{e}$$

To determine the force ratio we write the virtual work principle as

$$R(a\,\delta\psi) - f_g(b\,\delta\psi) = 0$$

which yields the relationship

$$f_g = \frac{a}{b}\,R$$

Those who doubt that the principle of virtual work holds advantages for some problems are urged to analyze this system by isolating free-bodies and writing corresponding equations of equilibrium.

Equivalence Between the Principle of Virtual Work and Vanishing Force Resultant.   Consider a system comprised of a collection of particles, the resultant force acting on the ith particle denoted by $\mathbf{f}_i$. It will prove useful to distinguish between external forces (those of interaction

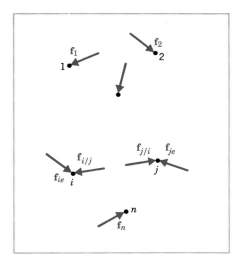

with bodies outside the system) and internal forces (those of interactions among particles within the system). We denote the external force acting on the ith particle by $\mathbf{f}_{ie}$ and the force exerted on the ith particle by the jth particle by $\mathbf{f}_{i/j}$. Then the condition that the ith particle be in equilibrium may be written as

$$\mathbf{f}_i = \mathbf{f}_{ie} + \sum_j \mathbf{f}_{i/j} = \mathbf{0} \qquad (7\text{-}3)$$

Now, let us dot multiply both sides of each such equation by a vector $\delta\mathbf{r}_i$ and add all the equations pertaining to the system:

$$\sum_i \mathbf{f}_{ie} \cdot \delta\mathbf{r}_i + \sum_i \sum_j \mathbf{f}_{i/j} \cdot \delta\mathbf{r}_i = 0$$

Newton's third law,

$$\mathbf{f}_{i/j} = -\mathbf{f}_{j/i}$$

permits simplification of the contribution to the double sum from the pair of terms

$$\mathbf{f}_{i/j} \cdot \delta\mathbf{r}_i + \mathbf{f}_{j/i} \cdot \delta\mathbf{r}_j = \mathbf{f}_{i/j} \cdot (\delta\mathbf{r}_i - \delta\mathbf{r}_j)$$

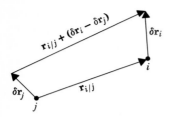

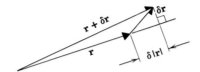

If the vectors $\delta\mathbf{r}_i$ and $\delta\mathbf{r}_j$ are small changes in position of the ith and jth particles, the vector $(\delta\mathbf{r}_i - \delta\mathbf{r}_j)$ is the change in the position vector $\mathbf{r}_{i/j}$ which locates the ith particle relative to the jth particle. Furthermore, we note that if $\delta\mathbf{r}$ is small compared to $\mathbf{r}$, the projection of $\delta\mathbf{r}$ onto $\mathbf{r}$ is equal to the change in the magnitude of $\mathbf{r}$; that is,

$$|\delta\mathbf{r}|\, \cos \sphericalangle {}^{\delta\mathbf{r}}_{\mathbf{r}} = \delta\, |\mathbf{r}|$$

Then, because $\mathbf{f}_{i/j}$ is parallel to $\mathbf{r}_{i/j}$,

$$\mathbf{f}_{i/j} \cdot \delta\mathbf{r}_{i/j} = |\mathbf{f}_{i/j}||\delta\mathbf{r}_{i/j}|\, \cos \sphericalangle {}^{\delta\mathbf{r}_{i/j}}_{\mathbf{r}_{i/j}}$$
$$= -T_{ij}\, \delta|\mathbf{r}_{i/j}|$$

where T_{ij} is defined as positive when the action-reaction is attractive and as negative when it is repulsive. With these results the equilibrium equation (7-3) can be rewritten as

$$\boxed{\delta W \equiv \sum_i \mathbf{f}_{ie} \cdot \delta\mathbf{r}_i - \sum_{i-j} T_{ij}\, \delta|\mathbf{r}_{i/j}| = 0}$$ (7-4)

in which the symbol $\displaystyle\sum_{i-j}$ means sum over all pairs of particles, and $\delta|\mathbf{r}_{i/j}|$ is the increase in the distance between the ith and jth particles.

The left-hand side of Equation 7-4 is the virtual work of the forces acting on the system. We see from the above derivation that $\delta W = 0$ follows from $\mathbf{f}_i = \mathbf{0}$, so that if the system is in equilibrium the virtual work necessarily vanishes. Furthermore, if Equation 7-4 is satisfied for *arbitrary* $\delta\mathbf{r}_i$, the equation $\delta W = 0$ implies that $\mathbf{f}_i = \mathbf{0}$; therefore Equation 7-4 is also a sufficient condition for equilibrium.

The meanings of the terms in Equation 7-4 may be reinforced by a reexamination of the examples presented earlier. In determining the relationship between P and M for the slider crank mechanism, we considered a virtual displacement in which the distances between all pairs of particles that interact remain constant; therefore, the virtual work of the internal forces, $\sum_{i-j} T_{ij}\,\delta|\mathbf{r}_{i/j}|$, is zero in this case. Of the external forces, the reaction from the wall against the piston is perpendicular to the virtual displacement of the piston so that this $\mathbf{f}_{ie} \cdot \delta\mathbf{r}_i$ is zero. And because the $\delta\mathbf{r}_i$ for the reaction at the crank support is zero, the virtual work of this external force is also zero. This leaves us with

$$\delta W = \sum_i \mathbf{f}_{ie} \cdot \delta\mathbf{r}_i$$
$$= -P\,\delta x - M\,\delta\theta$$

as written previously.

In determining the internal force in the connecting rod, we considered a different virtual displacement, this one in which the distance between the particles on either side of the separation undergoes a variation. That is, $\delta|\mathbf{r}_{i/j}| = \delta l$, and because these particles interact with a force R, the contribution to the virtual work there is

$$-\sum_{i-j} T_{ij}\,\delta|\mathbf{r}_{i/j}| = R\,\delta l$$

The only external force that would do work on this virtual displacement is P, so that

$$\sum_i \mathbf{r}_{ie} \cdot \delta\mathbf{r}_i = -P\,\delta x$$

In the analysis of the platform balance scale, the virtual displacement was chosen so that neither the internal forces nor the support reactions would do work. This illustrates the main strategy in using the principle of virtual work to best advantage: a virtual displacement that will involve as few unknown forces as possible in the equation of vanishing virtual work should be chosen; in this way one is spared the trouble of eliminating unknowns from a set of simultaneous equations.

Final Example

Determine the bending moment at the section where the force is applied to the beam in Figure 7-4a.

Consider the virtual displacement shown in Figure 7-4b. The virtual work that the bending moment would do on the left portion is equal to

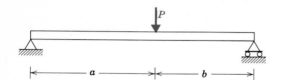

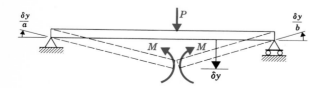

Figure 7-4

$-M\,\delta y/a$ and that which the corresponding reaction would do on the right portion is $-M\,\delta y/b$. The principle of virtual work then states

$$P\,\delta y - M\left(\delta\,\frac{y}{a} + \frac{\delta y}{b}\right) = 0$$

from which

$$M = \frac{ab\,P}{a+b}$$

The previous analysis which led to this result required that at least one of the support reactions be determined first.

Problems

7-6 From sketches of small displacements, determine the relationship

$$-\delta x = a\,(\sin\theta + \tan\phi\,\cos\theta)\delta\theta$$

for the virtual displacement shown in Figure 7-2*b*.

7-7 Use the principle of virtual work to determine the bending moment at the section located by x.

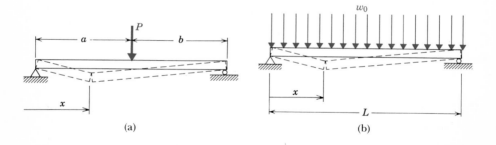

(a) (b)

7-8   Use the principle of virtual work to determine the force in member *AB* of the truss. Do *not* compute support reactions. All members are of equal length.

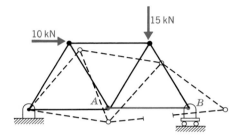

Use the method of virtual work to solve each of the following problems.

7-9   Problem 2-38.

7-10   Problem 4-11.

7-11   Problem 4-12.

7-12   Problem 4-13. Note that the connecting rod moves as a rotation around point *C'*, called the instantaneous center of rotation.

7-13   Problem 4-14.

7-14   Problem 4-15 (cable tension only).

7-15   The reaction at *A* in Problem 4-17.

7-16   The cable tension in Problem 4-19.

7-17   Problem 4-20.

7-18   The moment *M* in Problem 4-25.

7-19   The cable tension in Problem 4-26.

7-20   Problem 4-28.

7-21   The reaction at *C* in Problem 4-31(b).

7-22   Problem 5-7.

7-23   The force in member *BD* in Problem 5-14; determine the resultant of the two applied forces first.

7-24   Problem 5-33.

7-25   Problem 6-14.

APPENDIX A
SOME USEFUL
NUMERICAL VALUES

A-1

PHYSICAL CONSTANTS

Universal gravitational constant	6.67×10^{-11} N·m²/kg²
Mass of the earth	5.983×10^{24} kg
Speed of light	0.2998 Gm/s
Standard atmospheric pressure	101.325 kPa
Density of air (0°C, 1 atm)	1.29 kg/m³
Density of water (4°C, 1 atm)	1.0000 Mg/m³
Density of concrete	2.4 Mg/m³
Density of steel	7.7–7.9 Mg/m³

A-2

PREFIXES FOR SI UNITS

Exponents in higher order derived units apply to prefixes as well as units. For example, km² = (10^3 m)² = 10^6 m²

Prefix	SI Symbol	Multiplication Factor
tera	T	10^{12}
giga	G	10^9
mega	M	10^6
kilo	k	10^3
hecto*	h*	10^2
deka*	da*	10^1
deci*	d*	10^{-1}
centi*	c*	10^{-2}
milli	m	10^{-3}
micro	μ	10^{-6}
nano	n	10^{-9}
pico	p	10^{-12}
femto	f	10^{-15}
atto	a	10^{-18}

* Use discouraged.

A-3

UNITS OF MEASUREMENT

Quantity	SI Unit	Other Units
length	meter (m)	inch = 25.400 mm
		foot = 0.304 800 m
		statute mile = 1.609 344 km
		nautical mile = 1.852 000 km
		fathom = 1.828 800 m
mass	kilogram (kg)	pound-mass = 0.453 592 37 kg
		slug = 14.593 903 kg
		grain = 64.798 910 mg
time	second (s)	minute = 60 s
		hour = 3600 s
temperature	kelvin (K)	degrees Fahrenheit $t_K = (t_F + 459.67)/1.8$
		degreesCelsius $t_K = t_C + 273.15$
plane angle	radian (rad)	degree = $(\pi/180)$ rad
		minute = (1/60) deg
		second = (1/60) min
area	m²	acre = 4046.856 m²
		hectare = 10^4 m²
energy	joule (J)	foot-pound-force = 1.355 818 J
	$J = N \cdot m$	erg = 10^{-7} J
	$= kg \cdot m^2/s^2$	calorie (mean) = 4.190 02 J
		British thermal unit = 1.055 kJ
force	newton (N)	pound-force = 4.448 222 N
	$N = kg \cdot m/s^2$	kip = 4.448 222 kN
		poundal = 0.138 255 N
		dyne = 10^{-5} N
		kilogram-force = 9.806 650 N
power	watt (W)	horsepower (550 ft-lbf/s) = 745.7 W
	$W = J/s$	British thermal unit per hour = 0.293 071 W
	$= kg \cdot m^2/s^3$	refrigeration ton = 3.517 kW
pressure	pascal (Pa)	bar = 100 kPa
	$Pa = N/m^2$	pound per square inch = 6.894 757 kPa
	$= kg/m \cdot s^2$	centimeter of mercury (0°C) = 1.333 22 kPa
		centimeter of water (4°C) = 98.0638 Pa
velocity	m/s	kilometer per hour = 0.277 778 m/s
		mile per hour = 0.447 040 m/s
		knot = 0.514 444 m/s
volume	m³	liter = dm³ = 10^{-3} m³
		fluid ounce = 2.957 353 × 10^{-5} m³
		US liquid gallon = 3.785 412 × 10^{-3} m³
		barrel = 0.158 987 m³

APPENDIX B
PROPERTIES OF LINES, AREAS, AND VOLUMES

B-1

LINES

$L = \text{length}$ $c_i = \dfrac{1}{L} \displaystyle\int i \, dL$ $(i = x, y, z)$

$(c_x, c_y, c_z) = \text{coordinates of centroid}$

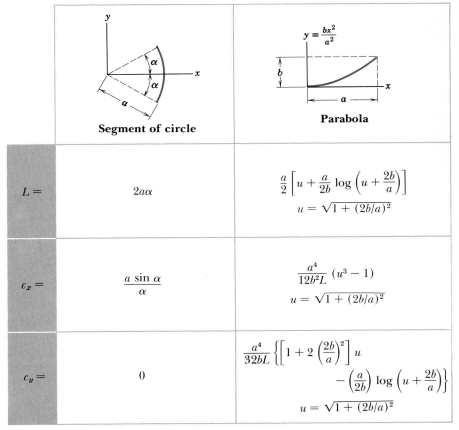

	Segment of circle	**Parabola** $y = \dfrac{bx^2}{a^2}$
$L =$	$2a\alpha$	$\dfrac{a}{2}\left[u + \dfrac{a}{2b}\log\left(u + \dfrac{2b}{a}\right)\right]$ $u = \sqrt{1 + (2b/a)^2}$
$c_x =$	$\dfrac{a\sin\alpha}{\alpha}$	$\dfrac{a^4}{12b^2L}(u^3 - 1)$ $u = \sqrt{1 + (2b/a)^2}$
$c_y =$	0	$\dfrac{a^4}{32bL}\left\{\left[1 + 2\left(\dfrac{2b}{a}\right)^2\right]u - \left(\dfrac{a}{2b}\right)\log\left(u + \dfrac{2b}{a}\right)\right\}$ $u = \sqrt{1 + (2b/a)^2}$

B-2

PLANE AREAS

A = Area $\quad c_i = \dfrac{1}{A} \int i \, dA \quad (i = x, y)$

$I_{xx} = \int y^2 \, dA \qquad I_{yy} = \int x^2 \, dA \qquad I_{xy} = -\int xy \, dA$

			$y = \dfrac{bx^n}{a^n}$, $n > -1$
$A =$	ab	$\dfrac{ab}{2}$	$\dfrac{ab}{n+1}$
$c_x =$	0	$\dfrac{a+c}{3}$	$\dfrac{n+1}{n+2}\, a$
$c_y =$	$\dfrac{b}{2}$	$\dfrac{b}{3}$	$\dfrac{(n+1)b}{2(2n+1)}$
$I_{xx} =$	$\dfrac{ab^3}{3}$	$\dfrac{ab^3}{12}$	$\dfrac{ab^3}{3(3n+1)}$
$I_{yy} =$	$\dfrac{a^3 b}{12}$	$\dfrac{ab}{12}(a^2 + ac + c^2)$	$\dfrac{a^3 b}{n+3}$
$I_{xy} =$	0	$-\dfrac{b^2 a}{24}(2c + a)$	$-\dfrac{a^2 b^2}{4(n+1)}$

	Quarter ellipse	Sector of circle	Segment of circle
$A =$	$\dfrac{\pi a b}{4}$	$a^2 \alpha$	$a^2 \left(\alpha - \dfrac{1}{2} \sin 2\alpha \right)$
$c_x =$	$\dfrac{4a}{3\pi}$	$\dfrac{2a \sin \alpha}{3\alpha}$	$\dfrac{4a}{3} \dfrac{\sin^3 \alpha}{2\alpha - \sin 2\alpha}$
$c_y =$	$\dfrac{4b}{3\pi}$	0	0
$I_{xx} =$	$\dfrac{\pi a b^3}{16}$	$\dfrac{a^4}{8} (2\alpha - \sin 2\alpha)$	$\dfrac{a^4}{8} \left(2\alpha - \dfrac{4}{3} \sin 2\alpha + \dfrac{1}{6} \sin 4\alpha \right)$
$I_{yy} =$	$\dfrac{\pi a^3 b}{16}$	$\dfrac{a^4}{8} (2\alpha + \sin 2\alpha)$	$\dfrac{a^4}{8} \left(2\alpha - \dfrac{1}{2} \sin 4\alpha \right)$
$I_{xy} =$	$-\dfrac{b^2 a^2}{8}$	0	0

B-3

VOLUMES

V = volume $\qquad c_i = \dfrac{1}{V} \displaystyle\int i\, dV \qquad (i = x,\, y,\, z)$

$(c_x,\, c_y,\, c_z)$ = coordinates of centroid

	Wedge	Segment of sphere
$V =$	$\dfrac{abc}{2}$	$\pi h^2 \left(a - \dfrac{h}{3} \right)$
$c_x =$	$\dfrac{2a}{3}$	$a\, \dfrac{(1 - h/2a)^2}{(1 - h/3a)}$
$c_y =$	$\dfrac{b}{3}$	0
$c_z =$	$\dfrac{c}{2}$	0

For the Segment of sphere: $h = a\,(1 - \cos \alpha)$

Cone

$V = \dfrac{Ah}{3}$

$c_x = \dfrac{3}{4}\, b_x$

$c_y = \dfrac{3}{4}\, b_y$

$c_z = \dfrac{1}{4}\, h$

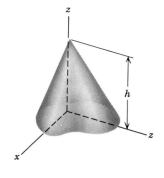

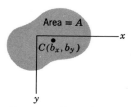

Area = A

$C(b_x, b_y)$

Base section of cone

REFERENCES

Metric Practice Guide (American National Standards Institute Booklet Z210.1, 1974).

Isaac Asimov, *Asimov's Biographical Encyclopedia of Science and Technology* (Doubleday and Company, 1972).

George W. Housner and D. E. Hudson, *Applied Mechanics Statics,* Second Edition (D. Van Nostrand Company, 1961).

Edward J. Routh, *Analytical Statics, Vol I* (Cambridge at the University Press, 1896.)

Murray R. Spiegel, *Vector Analysis* (Schaum's Outline Series, Schaum Publishing Company, 1959).

INDEX

Acceleration, 3
Acre, 190
Addition of vectors, 35
Angle, plane, 190
Archimedes' principle, 134
Area, 190
   centroid of, 124, 192-193
   first moment of, 124
   second moment of, 138, 192-193
Associative law
   of addition, 35
   of scalar multiplication, 38, 39
Atmospheric pressure, 129
Axes
   rectangular Cartesian, 41
   rotation of, 141
   translation of, 139
Axial force, 104
Axis, moment about, 48

Bar, 190
Barrel, 190
Beams
   bending moments in, 104, 148
   differential equations of equilibrium, 148
   shearing force in, 104, 148
Belt friction, 111
Bending moment, 104, 148
Bodies
   rigid, 20, 175
   submerged, 129
Bridge, suspension, 161
British thermal unit, 190
Btu, *see* British thermal unit
Buoyant force, 134

Cables
   catenary, 163
   differential equation of equilibrium, 160, 161, 162
   horizontally uniform load on, 161
   loaded under own weight, 162
   parabolic, 161
Cartesian components, 41
Catenary, 163
Center of gravity, 123
Center of mass, 122
Centroid, 124, 191-194
Clamped supports, 64, 88
Coefficient of friction, 108
Commutative law
   of addition, 35
   of cross multiplication, 39
   of scalar multiplication, 37
Composition of force systems, 46
Compressive force, 93
Concentrated couples on beams, 153
Concentrated forces on beams, 152
Conversion of units, 8

Coordinates, rectangular Cartesian, 41
Coulomb, C., 108
Coulomb friction, 108
Couple, 28
Couple-supporting members, 103
Cross product of vectors, 38

D'Alembert, 1
Deficient supports, 87
Derived quantities, 7
Diagram
  bending moment, 150, 152
  shearing force, 150, 152
Differential equation
  equilibrium, of a beam, 149, 150
    of a belt, 112
    of a cable, 160, 161, 162
Dimensional consistency, 9
Direction cosines, 44
Displacement, virtual, 178, 184
Distributive law
  of cross multiplication, 39
  of multiplication by a scalar, 37
  of scalar multiplication, 38
Dot product of vectors, 37
Dyne, 190

Energy, 190
Equilibrium, 59, 68, 184
Equivalent force systems, 29, 51, 121
Erg, 190

Fathom, 190
First moment of area, line, volume,
  124
Flexible cables, 159
Fluid element, 129
Fluid ounce, 190
Fluid pressure, 129
Foot, 190
Force, 2, 3, 190
  axial, 104
  buoyant, 134
  components of, 52
  distributed, 119
  internal, 62
  line of action, 21

moment of, 21, 46
resultant, 13
transmissibility of, 21
Force system
  composition of, 52
  equivalent, 51
  resolution of, 13, 46
Free-body diagram, 4, 60
Friction
  belt, 111
  coefficient of, 108
  Coulomb, 108
  dynamic coefficient of, 110
  limiting, 108
  static coefficient of, 110
Fundamental quantities, 7

Gallon, 190
Grain, 190
Graphical interpretations of load,
    shear, and bending moment,
    151
Gravitational constant, 5, 188
Gravity, 5
  center of, 123

Hamilton, 1
Hectare, 190
Horsepower, 190
Hour, 190

Improper support systems, 87
Inch, 190
Incipient motion, 108
Incompressible fluid, 131
Indeterminate support systems, 87
Inertia, moment of, 139, 192-193
Internal forces, 62
Isolation of free-body, 60-62

Joints, equations from, 93
Joule, 174, 190

Kilogram, 7, 190
Kilogram-force, 190
Kip, 190
Knot, 190

Lagrange, 1
Length, 7, 190
Limiting friction, 108
Line, centroid, 124, 191
Line of action, 20-21, 33
Liter, 190

Magnitude
  of a force, 4
  of a vector, 33
Mass, 3, 7, 11, 190
  center of, 122
Maximum moment of inertia, 143
Minimum moment of inertia, 144
Meter, 7, 190
Mile, 190
Minute, 190
Mohr, O., 142
Mohr's circle, 142
Moment
  bending, 104, 148
  twisting, 104
Moment of force
  about a point, 21, 46, 49
  about an axis, 48, 49
  resultant, 49
Moment of inertia of plane areas, 139,
    192-193
Multiplication
  by a scalar, 36
  cross, 38
  dot, 37
  scalar, 37
  vector, 38

Newton (unit of force), 3, 190
Newton, Isaac, 1
Newton's law of gravitation, 5
Newton's laws of motion, 2

Parabolic cable, 161
Parallel axis formulas, 140
Parallelogram law, 13
Pascal, 129, 190
Planar force systems, 60
Platform scale, 181
Position vector, 46

Pound, 11, 190
Poundal, 190
Pound-force, 190
Pound-mass, 190
Power, 190
Prefixes for SI units, 189
Pressure, 190
  static fluid, 129
Principle
  of transmissibility, 21
  of virtual work, 178, 183
Product
  dot, 37
  cross, 38
  scalar, 37
  vector, 38
Product of inertia of plane area, 139
Projected area interpretation, of com-
    ponent of fluid pressure force,
    131
Projection of a vector, 34

Radian, 190
Reaction, 2
Redundant supports, 87
Refrigeration ton, 8, 190
Resolution of force systems, 46
Resultant of forces, 13
Right-hand screw rule, 22, 38
Rigid body, 20
Rotation of axes, 141

Scalar, multiplication by, 36
Scalar product of vectors, 37
Scalar triple product, 45
Scale, platform, 181
Second, 7, 190
Second moments of area, 138, 192-
    193
Sections, equations from, 95
Shearing force, 104, 148
SI units of measurement, 8, 189-190
Sign convention for shearing force and
    bending moment, 149
Statically determinate and indetermi-
    nate support systems, 87
Slug, 11, 190

Submerged bodies, 129
Subtraction of vectors, 35
Supports, 64-66
    clamped, 88
    deficient, 87
    redundant, 87
    statically determinate and indetermi-
        nate, 87
Suspension bridge, 161

Temperature, 190
Tensile force, 93
Time, 7, 190
Transmissibility, 21
Trusses, 91
Twisting moment, 104

Unit vectors, 41
Units of measurement, 7, 190

Varignon, P., 51
Varigon's theorem, 24, 50

Vector
    addition, 35
    components, 41
    position, 46
    products, 36-39
    projection, 34
    subtraction, 35
    unit, 41
Velocity, 33, 190
Virtual displacement, 178, 184
Virtual work, 178, 183
Volume, 190, 194
    centroid of, 124, 194

Watt, 190
Weight, 11
Work done by a force, 173
Work
    done by a force, 173
    done on a rigid body, 175
    virtual, 178, 183
Wrench, 54

Some Units of Measurement

Systeme International

Quantity	Name	Abbre-viation	Basic Units	Other Equivalent
Mass	kilogram	kg	kg	0.068 52 sl 2.205 lbm
Length	meter	m	m	3.281 ft 39.370 in
Time	second	s	s	1.000 sec
Force	newton	N	kg·m/s²	0.2248 lbf
Energy	joule	J	kg·m²/s²	0.7376 ft-lbf
Power	watt	W	kg·m²/s³	0.001 341 HP
Pressure	pascal	Pa	kg/m·s²	0.000 1450 psi
Volume	cubic meter		m³	264.2 gal 35.31 ft³ 1000. liters

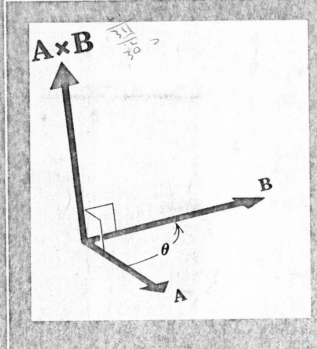

$$\mathbf{A} \cdot \mathbf{B} = AB \cos \theta.$$